T0205502

Advances in Nanotechnology and Its Applications

Ahmad Tariq Jameel · Abu Zahrim Yaser
Editors

Advances in Nanotechnology and Its Applications

Editors
Ahmad Tariq Jameel
Department of Biotechnology Engineering,
Faculty of Engineering
International Islamic University Malaysia
Kuala Lumpur, Malaysia

Abu Zahrim Yaser
Faculty of Engineering
Universiti Malaysia Sabah
Kota Kinabalu, Sabah, Malaysia

ISBN 978-981-15-4741-6 ISBN 978-981-15-4742-3 (eBook)
https://doi.org/10.1007/978-981-15-4742-3

This Springer imprint is published by the registered company Springer Nature Singapore Pte Ltd.
The registered company address is: 152 Beach Road, #21-01/04 Gateway East, Singapore 189721, Singapore

Preface

Advances in nanotechnology have led to the creation of new materials and devices with a vast range of applications in biomedicine, biomaterials, biosensors, nanoelectronics, energy production, food safety, consumer products, among many others. With new researches, there is rapidly increasing list of benefits and applications of nanotechnology. Growing popularity of nanotechnology depends on the fact that it is possible to tailor the structures of materials at extremely small scales to achieve specific properties. Nano-bioengineering of enzymes is aiming to enable conversion of cellulose into ethanol for fuel. Novel nanomaterials are being employed as support for enzyme immobilization to improve stability and efficiency of industrial biocatalysts. Cellulosic nanomaterials have demonstrated potential applications in a wide array of industrial sectors, including electronics, construction, packaging, food, energy, health care, automotive, and defense. Nanoscale materials are also being incorporated to improve performance into a variety of applications such as nano-structured ceramic coatings, nanotechnology-enabled lubricants, nanoparticles in catalysis, nano-engineered materials as household products, and in personal care products. The aim of this book is to bring forth to the audience some of the latest developments in nanotechnology, especially in the fields of engineered nanomaterials for enzyme immobilization for use in biocatalyst and biosensor technology, applications of nanomaterials in electronics, packaging materials industry, environmental technology, etc.

The opening chapter by Jameel and coauthors contributes to the important area of developing novel supports for enzyme immobilization. However, carbon-based nanomaterials (NMs) have gained high popularity among different immobilization support materials owing to its biocompatibility and large specific surface area. Immobilization technology is now increasingly used for enhancing enzyme stability and reuse. Chapter 1 starts with a comprehensive overview of enzyme immobilization on hydrogel/polymeric materials and onto nanoparticles/NMs followed by a detailed review of recent literature on the topic. Then, the authors discuss their research findings on the immobilization of β-glucosidase on multi-walled carbon nanotube (MWCNT) and alginate hydrogel. Kinetic studies reveal a higher enzyme affinity for the substrate immobilized onto MWCNT support compared to

Ca-alginate. It has been demonstrated that despite a general decrease in the enzyme activity due to immobilization, there is greater retention of activity and stability of the immobilized enzyme upon multiple cycles of hydrolysis. Overall, the chapter concludes that employing nanoparticles as support for enzyme immobilization bears distinct economic advantages for large-scale industrial applications.

Graphene owing to its biocompatibility, sensitivity, and fast electron transfer properties is increasingly being employed in electrochemical biosensor. In the second chapter, Sanober et al. have highlighted application of graphene to modify sensing electrodes for immune sensors that are based on the principle of antigen–antibody interaction. The chapter describes different techniques for the synthesis of graphene and its derivatives. Principles and different approaches of antibody immobilization on graphene-coated electrodes are discussed in detail. Pros and cons of each immobilization technique together with its methodology are surveyed.

There has been outpouring of the scientific literature on biosensor technology during last one decade or so. The third chapter by Shah and colleagues presents an extensive review of nanomaterials used in biosensor applications. Firstly, a brief account of the working principles and types of biosensors is provided and then its historical development is summarized. Nanomaterials having 1D, 2D, and 3D configuration and their application as biosensors are described. The working mechanism of biosensors is delineated in the light of the chemistry of biomolecule interaction with nano-structured materials. Biosensors based on different classes of materials are discussed in detail. Broadly, classification is made on: carbon-based nanomaterials covering carbon nanotube, graphene, and carbon nanofiber; metal-based nanoparticles such as functionalized metal nanoparticles and metal oxides; and nanocomposites encompassing hybrid polymer nanocomposites and metal oxide nanohybrid materials. The review specifically focusses on nanomaterials that give higher sensitivity, biocompatibility, improved selectivity and cost-effectiveness in biosensor performance.

The subsequent chapter by Akmal and Ahmad highlights the novel applications of piezoelectric biomaterials in the development of flexible electronic devices in energy harvesting, sensing, and biomedicine. The authors first introduce the very concept of piezoelectricity in crystalline materials and different classes of piezoelectric materials and their applications. The chapter focusses on the biomedical applications of thin film of piezoelectric bio-nanomaterials, as an alternative to the existing electronic health-monitoring devices. The physics of piezoelectricity is briefly but vividly explained. Biomaterials that exhibit piezoelectricity are natural polymers such as amino acid, cellulose, chitin, chitosan, and collagen, and these are described in detail. Different methods of quantifying piezoelectricity of biomaterials at the nanoscale are discussed. Prospects of potential applications such as nanogenerators, biosensors, vibration sensor, and other biomedical applications of piezoelectric nanomaterials are highlighted.

In view of the numerous applications of the nanoparticles (NPs), the synthesis of nanoparticles continues to attract much attention. Contribution of Soroodi et al. in the fifth chapter expounds green synthesis of nanoparticles from plant source as an alternative to chemical route of nanoparticle synthesis. The authors have described method to synthesize selenium (Se) nanoparticles using cocoa pod husk extract. The presence of biomolecules and phytochemicals in plants is believed to play an important role in the formation of stable nanoparticles. The review covers various plants used and effect of process parameters in the synthesis of nanoparticles. The effect of factors such as concentration of plant extract and salt solution, temperature, pH, etc., is known to influence the size distribution, morphology as well as the yield of the nanoparticles. The study surmised that the phytochemicals and enzymes present in the plant extract might be the primary source for the biosynthesis of Se NPs.

The sixth chapter introduces readers with nanocellulose, which can be prepared from any cellulose source material. Nanocelluloses discussed are of two types: cellulose nanofibril (CNF) or microfibrillated cellulose (MFC), and cellulose nanocrystal (CNC). Here, Jimat et al. describe the synthesis of cellulose nanofiber from cellulose obtained from cocoa pod husks (CPHs) and fibrillated with ultrasonication method. These nanocelluloses find novel applications owing to unique properties of low thermal expansion coefficient, outstanding reinforcing potential and transparency, low density, high aspect ratio, biocompatibility, high strength and stiffness. Especially CNC being amenable to chemical modifications such as fluorescent labeling find potential use in the field of biomedical applications such as biosensors, bioprobes, fluorescence bioassays, bioimaging, and so on. On the contrary, CNF owing to its characteristic morphology and physical properties is seen as a promising material for applications such as filter material, and high gas barrier packaging material. Besides, it finds use in electronic devices, food, medicine, cosmetics, and health care products. The authors have described in detail the experimental methodology and analytical procedures for the synthesis and characterization of nano-structured celluloses.

The seventh chapter extends the scope of nanotechnology to the development of new packaging materials. Conventional packaging materials are mostly non-biodegradable and pose tremendous environmental problems. In quest for biodegradable packaging materials, polylactic acid (PLA), an aliphatic polyester, has been a topic of research lately. Ali et al. studied the synthesis and characterization of polylactic acid/organoclay nanocomposites as a substitute for conventional packaging materials. The agglomeration of organoclay with PLA exhibited improved modulus of the PLA/organoclay nanocomposites.

Finally, the last chapter reviews the potential application of zeolite and chitosan as adsorbents for gross pollutant traps (GPTs) in various water bodies such as storm water runoff and engineered conveyance. Safie et al. have briefly surveyed the GPTs performance in the removal of dissolved pollutants, COD and BOD, as well as the efficacy of zeolite and chitosan as adsorbent media. Though the subject of this chapter does not directly fits in the realm of nanotechnology, zeolite and chitosan can treat dissolved contaminants including those of ionic species by adsorption and ion exchange which are primarily nanoscale processes.

The current book presents different aspects of recent developments and progress in the synthesis and characterization of novel nanomaterials to tackle problems arising in bioprocess industry, biomedicine, and environment. The timely publication of this book hopefully would be of interest to students, teachers, and practitioners alike in the field of nanotechnology. The subject of nanotechnology is expected to remain at the forefront of cutting-edge research in science and technology throughout the next decades given the increasing integration of nanomaterials in diverse human activities.

I wish to express my deepest gratitude to all of the authors who contributed to this book with pertinent review articles and research chapters on different aspects of nanomaterial development and uses. I hope that the meticulous effort and the long hours of work invested by the authors and editors would prove worthy of reader's interest. Further, constructive suggestions of reviewers toward improving the quality of the manuscript are gratefully acknowledged. It is my pleasure to express my deep appreciation to Dr. Abu Zahrim Yaser for providing editorial support in general and keeping close liaison with the publishers during the preparation of the book. Further, I would like to thank Ms. Megana Dinesh, Mr. Viju Falgon and rest of the editorial staff at Springer Nature Singapore, for their full cooperation throughout the production process.

Kuala Lumpur, Malaysia
March 2020

Ahmad Tariq Jameel

Contents

Editors and Contributors

About the Editors

Ahmad Tariq Jameel is Associate Professor of chemical/biochemical engineering at the International Islamic University Malaysia (IIUM). He obtained his Ph.D. in chemical engineering from the Indian Institute of Technology Kanpur. Prior to joining IIUM, he had served as full-time faculty at universities in Saudi Arabia, Oman, Malaysia, and India. His current research interest is directed toward the enzyme-based biosensors and design of immobilized catalysts, among others. He has published over 80 research papers in international and national journals and conference proceedings. Besides, he has several chapters and edited books to his credit. He is Member of the *International Association of Colloid and Interface Scientists (IACIS)* and the *Canadian Society for Chemical Engineering (CSChE).*

Abu Zahrim Yaser is Associate Professor of waste processing technology at Universiti Malaysia Sabah (UMS). He obtained his Ph.D. from Swansea University. He has published 4 books, 16 chapters, 34 journals, and 50++ other publications. He was Guest Editor for *Environmental Science and Pollution Research* special issue (Springer). The Elsevier (UK) has recognized him as the Outstanding Reviewer for the Journal of Environmental Chemical Engineering. He was Visiting Scientist at the University of Hull and Member of Board of Engineers (Malaysia) and MyBIOGAS.

Contributors

Farah Binti Ahmad Department of Biotechnology Engineering, Faculty of Engineering, International Islamic University Malaysia, Kuala Lumpur, Malaysia

Mohd Hatta Maziati Akmal Department of Science in Engineering, Faculty of Engineering, International Islamic University Malaysia, Kuala Lumpur, Malaysia

Fathilah Binti Ali Department of Biotechnology Engineering, Kulliyyah of Engineering, International Islamic University Malaysia, Kuala Lumpur, Malaysia

S. M. Anisuzzaman Chemical Engineering Programme, Faculty of Engineering, Universiti Malaysia Sabah, Jalan UMS, Kota Kinabalu, Malaysia

Hazleen Anuar Department of Manufacturing and Materials Engineering, Kulliyyah of Engineering, International Islamic University Malaysia, Kuala Lumpur, Malaysia

Azlin Suhaida Azmi Department of Biotechnology Engineering, Kulliyyah of Engineering, International Islamic University Malaysia, Kuala Lumpur, Malaysia

Piravin Raj Barthasarathy Department of Biotechnology Engineering, Faculty of Engineering, International Islamic University Malaysia, Gombak, Kuala Lumpur, Malaysia

Parveen Jamal Department of Biotechnology Engineering, Kulliyyah of Engineering, International Islamic University Malaysia, Kuala Lumpur, Malaysia

Jamarosliza Jamaluddin Department of Bioprocess and Polymer Engineering, Faculty of Chemical and Energy Engineering, Universiti Teknologi Malaysia, Johor Bahru, Malaysia

Ahmad Tariq Jameel Department of Biotechnology Engineering, Faculty of Engineering, International Islamic University Malaysia, Gombak, Kuala Lumpur, Malaysia

Dzun Noraini Jimat Department of Biotechnology Engineering, Kulliyyah of Engineering, International Islamic University Malaysia, Kuala Lumpur, Malaysia

Mohammad Khalid Graphene and Advanced 2D Materials Research Group (GAMRG), School of Science and Technology, Sunway University, Petaling Jaya, Selangor, Malaysia

Labiba Mahmud Department of Biotechnology Engineering, Faculty of Engineering, International Islamic University Malaysia, Gombak, Kuala Lumpur, Malaysia

Mohd Zulhisham Moktar Chemical Engineering Programme, Faculty of Engineering, Universiti Malaysia Sabah, Jalan UMS, Kota Kinabalu, Malaysia

Nabisab Mujawar Mubarak Department of Chemical Engineering, Faculty of Engineering and Science, Curtin University Malaysia, Miri, Sarawak, Malaysia

Wan Mohd Fazli Wan Nawawi Department of Biotechnology Engineering, Kulliyyah of Engineering, International Islamic University Malaysia, Kuala Lumpur, Malaysia

Ibrahim Ali Noorbatcha Department of Biotechnology Engineering, Kulliyyah of Engineering, International Islamic University Malaysia, Kuala Lumpur, Malaysia

Sharifah Shahira Syed Putra Department of Biotechnology Engineering, Kulliyyah of Engineering, International Islamic University Malaysia, Kuala Lumpur, Malaysia

Mariani Rajin Chemical Engineering Programme, Faculty of Engineering, Universiti Malaysia Sabah, Jalan UMS, Kota Kinabalu, Malaysia

Mohd Hazman Saafie Chemical Engineering Programme, Faculty of Engineering, Universiti Malaysia Sabah, Jalan UMS, Kota Kinabalu, Malaysia

Nurliyana Nasuha Safie Chemical Engineering Programme, Faculty of Engineering, Universiti Malaysia Sabah, Jalan UMS, Kota Kinabalu, Malaysia

Syed Tawab Shah Graphene and Advanced 2D Materials Research Group (GAMRG), School of Science and Technology, Sunway University, Petaling Jaya, Selangor, Malaysia

Ihda Uswatun Shalihah Shohibuddin Department of Biotechnology Engineering, Faculty of Engineering, International Islamic University Malaysia, Gombak, Kuala Lumpur, Malaysia

Fatemeh Soroodi Department of Biotechnology Engineering, Kulliyyah of Engineering, International Islamic University Malaysia, Kuala Lumpur, Malaysia

Rashmi Walvekar Department of Chemical Engineering, School of Energy and Chemical Engineering, Xiamen University Malaysia, Sepang, Selangor, Malaysia

Wan Wardatul Amani Wan Salim Department of Biotechnology Engineering, Faculty of Engineering, International Islamic University Malaysia, 50728 Gombak, Kuala Lumpur, Malaysia

Abu Zahrim Yaser Chemical Engineering Programme, Faculty of Engineering, Universiti Malaysia Sabah, Jalan UMS, Kota Kinabalu, Malaysia

Faridah Yusof Department of Biotechnology Engineering, Faculty of Engineering, International Islamic University Malaysia, Gombak, Kuala Lumpur, Malaysia

List of Reviewers

Dr. Danish Mohammed, Universiti Kuala Lumpur, Kuala Lumpur, Malaysia
Dr. Dong Jin Kang, Leibniz Institute for New Materials, Saarbrücken, Germany
Dr. Faizah bt. Mohd Yasin, Universiti Putra Malaysia, Serdang, Selangor, Malaysia
Dr. Golnoush Zamri, University of Malaya, Kuala Lumpur, Malaysia
Dr. Khairatun Najwa Mohd Amin, Universiti Malaysia Pahang, Kuantan, Pahang, Malaysia
Dr. Lee Hooi Ling, Universiti Sains Malaysia, Penang, Malaysia
Prof. Mamoun Bader, Al Faisal University, KSA
Prof. Mohammed Saedi Jami, International Islamic University Malaysia, Kuala Lumpur, Malaysia
Prof. Mohammad Khalid, Sunway University, Bandar Sunway, Selangor, Malaysia
Dr. Mubarak Mujawar, Curtin University Malaysia, Miri, Sarawak, Malaysia
Dr. Shahid Mehmood, Fudan University, Shanghai, China
Prof. Zahangir Alam, International Islamic University Malaysia, Kuala Lumpur, Malaysia
Dr. Zurina Binti Mohamad, Universiti Teknologi Malaysia, Skudai, Johor Bahru, Malaysia.

Characterization of Enzyme Immobilization on Novel Supports—Multi-walled Carbon Nanotube and Alginate

Ahmad Tariq Jameel, Labiba Mahmud, and Faridah Yusof

Abstract Enzymes are preferred over chemical catalysts in a myriad of applications owing to their high specificity, selectivity and moderate operating conditions. Enzymes in the soluble state are susceptible to instability and difficult in separation. Immobilization of enzyme onto a support increases its physical and thermal stability, reusability and recovery from the reaction broth. Different support materials such as polymers, hydrogels, nanoparticles, nanofibers and nano-scaffolds are being used for enzyme immobilization. Carbon-based nanomaterials have gained high popularity among different support materials. Here, we present the research findings on the immobilization of β-glucosidase onto two novel support materials, i.e., glutaraldehyde-activated multi-walled carbon nanotubes (MWCNTs) and Ca-alginate beads. The relative merits of the two supports are compared in terms of the performance of the enzyme in each case. β-Glucosidase immobilized on glutaraldehyde-modified MWCNTs exhibited higher residual activity and stability compared to the enzyme encapsulated in Ca-alginate beads. Kinetic study shows a higher enzyme affinity for the substrate for enzyme immobilized onto MWCNT support compared to Ca-alginate. The overall results demonstrate that despite a general decrease in the enzyme activity due to immobilization, there is greater retention of activity of the immobilized enzyme upon multiple cycles of hydrolysis. This study provides distinct economic advantage of employing nanoparticles as support for enzyme immobilization for large-scale industrial applications.

Keywords Enzyme activity · Immobilization · β-Glucosidase · Carbon nanotube · Alginate · Kinetics · Reusability · Desorption

---

A. T. Jameel (✉) · L. Mahmud · F. Yusof
Department of Biotechnology Engineering, Faculty of Engineering, International Islamic University Malaysia, 50728 Gombak, Kuala Lumpur, Malaysia
e-mail: atjameel@yahoo.com

A. T. Jameel and A. Z. Yaser (eds.), *Advances in Nanotechnology and Its Applications*, https://doi.org/10.1007/978-981-15-4742-3_1

1 Introduction

Enzymes are preferred over chemical catalysts in pharmaceutical and bioanalytical applications, and food industries owing to their high specificity and selectivity as well as their moderate operating conditions. The enzyme β-glucosidase is widely used for saccharification of lignocellulosic biomass at commercial scale [1–3] and has also been proposed as an indicator for changes in soil quality [4]. β-Glucosidase is an important component of cellulase enzyme complex that is responsible for the hydrolysis of cellulose into glucose. Cellulose, being the major organic compound in plant biomass, there is great demand for efficient conversion of cellulose to bioenergy products to meet the increasing demand for bioenergy worldwide. Bacterial β-glucosidase is a rate-limiting factor during enzymatic hydrolysis of cellulose [5]. Thus, β-glucosidase enzyme is of great interest in biofuel industry for the hydrolysis of lignocellulosic biomass as a precursor for the production of biofuels, such as second-generation ethanol [6]. Some strains of β-glucosidase (e.g., EC 3.2.1.21) from living organisms are known to catalyze the hydrolysis of various glycosides that release free aroma constituent [7]. Most of the free or naturally occurring enzymes are susceptible to denaturation, under adverse operating conditions and storage. Moreover, enzymes in soluble state are not recoverable in active form for reuse and may contaminate the product formed [8]. Immobilization has been most successful in overcoming these limitations and allows greater control over process parameters as well as multiple usage of the enzyme. Immobilization of an enzyme refers to confinement of an enzyme in a certain region of a support material with retention of its activity so that it can be used repeatedly without experiencing much loss of its activity. Immobilization of enzyme onto a matrix increases its physical and thermal stability, reusability and recovery from the reaction broth. The immobilization techniques involve binding of an enzyme to a support material, which is usually activated by a cross-linker that attaches to suitable binding sites on the enzyme as shown in Fig. 1 [9]. Glutaraldehyde (GA) has been widely used as a cross-linker for a wide range of immobilization support materials such as hydrogel matrix and nanomaterials. However, other activating agents such as carbodiimide have also been used for functionalization of MWCNTs to covalently link the cellulose [10]. The common techniques to bind enzyme to the support matrix are: entrapment, cross-linking,

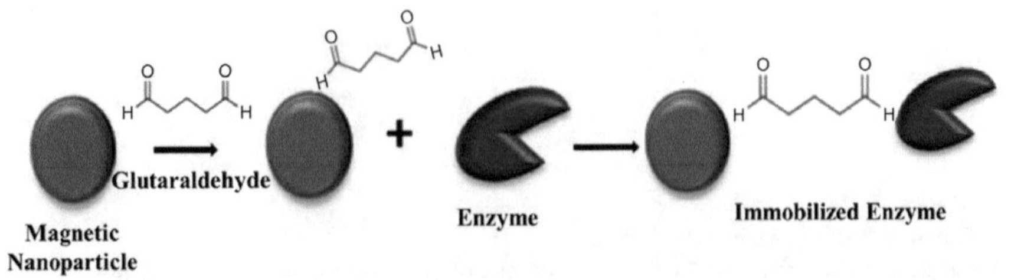

Magnetic Nanoparticle — Glutaraldehyde → + Enzyme → Immobilized Enzyme

Fig. 1 Schematic representation of binding of an enzyme with the glutaraldehyde-activated nanoparticle. GA acts as a cross-linking agent. Reproduced from Abraham and Puri [9]

adsorption and encapsulation. Different support materials have been used for enzyme immobilization such as hydrogels, e.g., alginate, carrageenan, etc.; nanoparticles, e.g., CNT, graphene, magnetic nanoparticles, etc.; and nanofibers.

Progress in nanotechnology over past two decades has led to development of novel nano-scaffolds that can be used as support for enzyme immobilization. As a result, nano-biocatalysis that integrates the biocatalyst and nanoscale materials has emerged as innovative technology in industrial bioprocesses. The obvious advantages of using nanoscale structures for enzyme immobilization are the reduced diffusional limitations and increased enzyme loading owing to high functional surface area. Among the various nanomaterials (NMs) being used as immobilization support, carbon-based NMs have gained high popularity due to their large specific surface area, biocompatibility and amenability to surface modification by functionalization, thereby resulting into efficient enzyme loading [11].

While the immobilization techniques involving entrapment or encapsulation such as in the case of alginate beads have been well documented and have shown great versatility [12], however they have shown poor performance on reusability due to enzyme leakage from the gel beads [13]. The increasing diversity in enzyme applications has led to a continuous search of new and improved methods for immobilization. Biocatalytic performance of immobilized enzyme largely depends on the enzyme and the support characteristics, as well as immobilization technique and process conditions employed [9].

1.1 Enzyme Immobilization in Hydrogel/Polymeric Materials

Hydrogel matrices and polymeric materials have been widely employed as support materials for immobilization of enzymes. In an earlier work, cross-linked β-glucosidase from *Aspergillus niger* was immobilized in calcium alginate. The immobilized enzyme exhibited changed activity behavior with temperature and pH, in comparison with the free enzyme. However, optimum pH and temperature were not altered upon immobilization. The thermal stability of the enzyme showed improvement after immobilization [14].

Immobilized β-D-glucosidase (isolated from *Pseudomonas pickettii* and *Aspergillus niger*) in calcium alginate beads by entrapment demonstrated about 15–26% of residual activity upon multiple uses. The optimum pH for maximum activity of the immobilized bacterial and fungal enzymes was reported as 5.0 and 3.0, respectively, values close to those of the free enzymes. Kinetic behavior of bacterial β-glucosidase exhibited substrate inhibition, a deviation from the Michaelis kinetics, and showed similar K_m values for immobilized and soluble enzymes. The K_m of the immobilized fungal enzyme was larger than that of the free enzyme, possibly due to increased resistance to substrate and product diffusion through the alginate network. The V_{max} values of immobilized β-glucosidases from *P. pickettii* and *A. niger* were 29.6 and 14.7 mM/min, respectively, smaller than those of the native enzymes [15].

Immobilization of β-glucosidase on alginate by the method of cross-linking–entrapment–cross-linking retained 46% residual activity under optimum conditions. Immobilization did improve the thermal and pH stabilities of β-glucosidase to some degree. The K_m value for the immobilized enzyme was estimated to be 1.97 mM. Immobilization allowed a significant improvement in storage stability and reusability of the enzyme. The immobilized β-glucosidase was found effective in enhancing the aroma of tea beverage [16].

β-Glucosidase immobilized on magnetic chitosan microspheres was investigated for the hydrolysis of cellulosic biomass using cellobiose as the model substrate [17]. The immobilized enzyme displayed an activity of 6.4 U/g support under optimum conditions. Immobilization resulted in the marginal increase of K_m, low shift in the optimal pH and improved thermal stability relative to the free enzyme. The immobilized β-glucosidase when applied to enzymatic hydrolysis of corn straw was able to maintain conversion rate above 76.5% after eight batches of reuse, depicting high operational stability of the immobilized enzyme [17].

Tan and Lee [18] studied the immobilization of β-glucosidase on glutaraldehyde-activated κ-carrageenan–polyethyleneimine (PEI) polyelectrolyte complex beads through cross-linking method. The biocatalyst displayed an immobilization yield of 98.4% and enzyme activity of 36 U/g. Immobilization led to an improved pH and thermal stability of β-glucosidase [18]. Another promising application of immobilized β-glucosidase in biotechnological industry is demonstrated for the production of resveratrol by Zhang et al. [19]. The enzyme was immobilized on cross-linked chitosan microspheres modified by L-lysine that retained up to 93% enzyme activity after continuous hydrolysis for 8 h. Silva et al. [20] studied immobilization of β-glucosidase from *Aspergillus japonicus* on aminoethyl-*N*-aminoethyl (MANAE)-agarose and DEAE-cellulose by ionic interaction. The MANAE and DEAE derivatives retained 50 and 60% residual activity respectively, after 5 cycles of reaction with the substrate pNP-β-D-glucopyranoside. The K_m and V_{max} values were also obtained in the study.

Normally, leaching of immobilized enzyme from the gel beads occurs due to the fact that the enzyme molecule has much smaller size than the pore size of the gel beads (~200 nm). To overcome this problem, a cross-linked cellulase aggregate (XCA) having colloidal size (~300 nm) was immobilized in calcium alginate beads to hydrolyze cellulose substrates [21]. Almost 96.4% of the XCA was retained inside the alginate beads after 10 cycles of hydrolysis of microcrystalline cellulose. The XCA-alginate biocatalyst retained 67% of the original activity after 10 cycles in contrast to complete loss of activity for the free cellulase immobilized in the alginate beads. Similar strategy can also be applied to other types of cross-linked enzyme aggregates to minimize enzyme leaching [21]. A thermo-responsive polymer (P_{NMN}) was developed as a novel carrier to immobilize cellulose as bioconjugate (P_{NMN}-C) that could be dissolved and recovered as precipitates [22].

1.2 Nanomaterials as Enzyme Immobilization Support

Nanomaterials provide a wider range in balancing the key factors which determine enzyme efficiency including specific surface area, mass transfer resistance, thermal and pH stability and effective enzyme loading [7, 9, 10, 23–28]. Nanoparticle-based immobilization allows nanoparticle–enzyme complex to be easily synthesized without toxic reagents, forming homogenous nanoparticles with thick enzyme shells as well as providing the opportunity of particle size tailoring [26]. Use of magnetic nanoparticles (MNPs) for enzyme immobilization can greatly facilitate the recovery of catalyst from the reaction medium as they can be easily separated by applying an external magnetic field [29, 30]. The immobilized cellulase onto Fe_3O_4 magnetic nanoparticles via glutaraldehyde activation showed enhanced stability and activity compared to the free enzyme [31].

In another study, β-glucosidase from *Aspergillus niger* was immobilized to functionalized magnetic nanoparticles by covalent binding that showed 93% of the immobilized enzyme successfully attached to the nanoparticles [25]. Using cellobiose as model substrate, the activity, thermal stability and recyclability of free and immobilized enzyme were characterized. Immobilized enzyme exhibited enhanced pH and thermal stability.

β-Glucosidase covalently immobilized on glutaraldehyde-activated chitosan–MWCNT composite carrier achieved optimum immobilization yield of 95.22%. Optimum pH for free and immobilized enzyme was found to be 6.0 and 5.0, respectively. Optimum temperature corresponding to maximum enzyme activity was increased from 35 to 45 °C after immobilization. Immobilization leads to better pH, thermal and storage stability than the free enzyme. The immobilized enzyme retained 68.4% of the initial activity after 50 days of storage at 4 °C. Immobilized enzyme upon ten cycles of multiple uses could retain as high as 72.83% of its initial activity. The kinetic constants, K_m and V_{max} for immobilized β-glucosidase, were found as 5.55 mM and 7.14 U/mg protein, respectively [7].

Purified β-glucosidase from *S. griseus* was successfully immobilized onto zinc oxide (ZnO) nanoparticles by adsorption. The immobilized enzyme exhibited greater thermostability and higher optimum temperature than the free enzyme [28]. Coutinho et al. [32] demonstrated efficient immobilization of β-glucosidase by adsorption onto hydroxyapatite nanoparticles. The immobilized enzyme displayed high enzyme binding capacity (around 50 mg protein g^{-1} support) and immobilization yield, with high enzyme affinity for the support over wide ranges of pH and ionic strength, and over 90% recovery of the activity.

In the present study, we investigated the immobilization of β-glucosidase (EC 3.2.1.21) from sweet almonds on two different classes of support materials, i.e., hydrogel matrix and nanoparticles. The enzyme was immobilized: (1) by entrapment in Ca-alginate beads and (2) onto the glutaraldehyde-activated multi-walled carbon nanotubes (MWCNTs) by covalent bonding. Hence, a comparative study was made of the two above-mentioned immobilization techniques by comparing the activities of the immobilized β-glucosidase to that of its native form. MWCNT was activated using glutaraldehyde which binds the enzyme to the nanoparticle surface allowing reagents to reach catalytic site while preventing contact with surrounding medium [24, 33–35]. MWCNT modified by glutaraldehyde was analyzed by FTIR spectroscopy. The effect of immobilization on the surface morphology of the MWCNT support was studied by SEM imaging.

The hydrolysis of the substrate *p*-nitrophenyl-β-D-glucopyranoside (pNPG) to p-nitrophenol (pNP) in the presence of β-glucosidase was studied for different substrate concentrations at pH 4.8 and 50 °C [1]. The enzyme activity was determined by calculating the concentration of *p*-nitrophenol using spectroscopic method. The results demonstrated that despite a general decrease in the activity due to immobilization, there was greater retention of activity upon reuse of the immobilized enzyme. β-Glucosidase immobilized on glutaraldehyde-functionalized MWCNTs exhibited higher residual activity compared to the encapsulated enzyme in Ca-alginate beads [36, 37].

2 Experimental Procedure for Enzyme Immobilization and Its Characterization

β-Glucosidase from sweet almonds (EC 3.2.1.21) and *p*-nitro phenyl-β-D-glucopyranoside (pNPG) were obtained from Sigma-Aldrich, sodium alginate and glutaraldehyde were obtained from Fisher Chemical, and calcium chloride was purchased from Merck. MWCNT was obtained from Sab Bayan Enterprise (Malaysia). All other chemicals used were commercially available reagent grade. In what follows, a brief description of the experimental procedure is given, however, the detailed methodology and experimental results may be found elsewhere [36].

2.1 Scanning Electron Microscope (SEM) Imaging of MWCNT

The micrographs of the MWCNT samples under different magnifications were obtained using scanning electron microscope (model JSM-5600). The imaging prior to functionalization and after immobilization was carried out on dry MWCNT sample. After the activation of MWCNTs with glutaraldehyde and subsequent immobilization with β-glucosidase, MWCNTs were freeze-dried and then observed under SEM for post-immobilization study [36].

2.2 FTIR of MWCNT

The Fourier transform infrared (FTIR) spectroscopy is generally employed to identify chemicals/chemical structure on the basis of the absorbance spectrum of the substance by measuring absorption of the light emanating from an infrared light source. The test sample is first exposed to different wavelengths of infrared light, and the device measures the wavelengths that are absorbed to generate a readable absorbance spectrum using Fourier transform method. This spectrum can then be compared to a library of spectra to find a match. In order to observe the changes in the surface morphology of the functionalized MWCNT, FTIR was performed on two sets of samples: the pristine MWCNT, before any treatment; and the functionalized MWCNT after glutaraldehyde modification and freeze-drying. In both samples, potassium bromide (KBr) was used as beam splitter.

2.3 Enzyme Preparation

In order to prepare β-glucosidase solution for each run of enzymatic reaction, 1.05 mg of lyophilized enzyme was dissolved in 2 mL of 0.1 M potassium phosphate buffer, at pH 6.4. The original (native) enzyme activity was measured as 9.46 U/mg.

2.4 Enzyme Immobilization in Ca-Alginate Beads

Enzyme solution (225 U) and sodium alginate solution were mixed in appropriate ratio to give a 4% (w/v) sodium alginate in the final mixture containing 40 μL of enzyme in 960 μL of sodium alginate solution [1, 12]. The mixture thus obtained was pipetted into 0.2 M calcium chloride solution under continuous stirring condition using 1-mL pipette from a height of approximately 20 cm for uniform bead formation. The beads were then incubated for 2 h in the calcium chloride solution that resulted into approximately 3-mm diameter beads. The calcium alginate beads immobilized with β-glucosidase were then thoroughly washed with distilled water twice and stored at 4 °C for further studies.

2.5 Surface Functionalization of MWCNT

MWCNT was washed with deionized water and then recovered by centrifuging for 10 min at 2000 rpm. Washed MWCNT was next suspended in 1 M glutaraldehyde in a shaker at 120 rpm for 6 h. The activated MWCNT was then removed by centrifugation, washed twice with deionized water to remove traces of glutaraldehyde, subsequently washed with 100 mM potassium phosphate buffer (pH 6.4) and used for further studies [34, 36].

2.6 Adsorption of β-Glucosidase on MWCNT

Modified MWCNT (0.1 g) was added to a mixture of β-glucosidase (225 U) and potassium phosphate buffer solution (pH 6.4) and kept overnight on shaker at 30 °C for immobilization. It has been suggested that even though immobilization may be rapid, a longer reaction time will render better results and hence the choice of overnight immobilization was made [37]. Immobilized β-glucosidase was separated by centrifugation at 2000 rpm for 20 min. This MWCNT-bound β-glucosidase was then washed three times with 100 mM potassium phosphate buffer (pH 6.4). The immobilized enzyme was finally stored in fresh buffer at 4 °C for further use [34, 36].

2.7 Enzyme Activity Assay

The enzymatic activity of β-glucosidase was determined using p-nitrophenyl-β-D-glucopyranoside (pNPG) as the substrate. The reaction mixture, containing 1.8 mL of 8 mM pNPG solution in 0.05 M sodium citrate buffer having pH 4.8, and 1.7 mL of calcium alginate beads were incubated at 50 °C in a water bath. After the enzymatic reaction had proceeded for 20 min, 0.5 mL of 1.0 M sodium carbonate was added to stop the reaction and develop the color of any *p*-nitrophenol (generating *p*-nitrophenolate) that has been released from the substrate. The absorbance was read at 410 nm, and the amount of pNP liberated was determined using standard graph. In case of the enzyme immobilized onto functionalized MWCNT, a similar protocol was followed where 1.7 mL of the glutaraldehyde-modified and β-glucosidase-bound MWCNT was used. The free enzyme assay consisting of enzyme dissolved in 1.7 mL of potassium phosphate buffer was used to determine activity of native β-glucosidase following above protocol. A unit of enzyme activity was defined as the amount of enzyme required to liberate 1 μmol of p-nitrophenol (pNP) per min from the hydrolysis of substrate under assay condition [1, 38, 39].

2.8 Determination of Kinetic Parameters

The kinetics of the unbound and immobilized β-glucosidase was investigated following the procedure described in Doran [40]. Activity assay protocol as described in Sect. 2.7 was used to carry out enzymatic reaction at 50 °C, for different initial substrate (pNPP) concentrations in the range of 0.5–8 mM. The product (pNP) concentration calculated using spectroscopic method was used to calculate the rate of reaction. Finally, the rate data were used to plot the Michaelis–Menten equation as well as different linear transformations of the Michaelis–Menten equations described below, in order to evaluate the best kinetic constants. Michaelis–Menten equation and its linear transformations are expressed as [40]:

$$\text{Michaelis–Menten equation:} \quad V = \frac{V_{\text{max}}[S]}{K_{\text{m}} + [S]} \tag{1}$$

$$\text{Lineweaver–Burk:} \quad \frac{1}{V} = \frac{K_{\text{m}}}{V_{\text{max}}}\frac{1}{S} + \frac{1}{V_{\text{max}}} \tag{2}$$

$$\text{Hanes(Langmuir)}: \quad \frac{S}{V} = \frac{K_{\text{m}}}{V_{\text{max}}} + \frac{S}{V_{\text{max}}} \tag{3}$$

$$\text{Eadie–Hofstee:} \quad \frac{V}{S} = \frac{V_{\text{max}}}{K_{\text{m}}} - \frac{V}{K_{\text{m}}} \tag{4}$$

where V and S are the volumetric rate of reaction and the substrate concentration, respectively. K_{m} and V_{max} are Michaelis constant and maximum reaction velocity, respectively. K_{m} is defined as the substrate concentration corresponding to rate of reaction equal to half of the maximum reaction velocity, $V_{\text{max}}/2$ [40]. The kinetic constants K_{m} and V_{max} were calculated using Lineweaver–Burk, Hanes, Eadie–Hofstee and hyperbolic regression (Michaelis–Menten) plots. At times, these linear transformations of the rate data can result in artificial weighting of the data leading to erroneous estimates of V_{max} and K_{m}. Thus, R^2 values from linear plots in conjunction with the predictions of the hyperbolic regression plot can be used as guiding criterion to identify the most accurate values of the kinetic constants.

2.9 *Reusability of Immobilized β-Glucosidase*

The activity of immobilized enzyme was calculated after each successive hydrolytic assay to assess the reusability of the enzyme upon multiple usages. After the first activity assay, the immobilized β-glucosidase on the MWCNT was recovered by centrifugation at 4000 rpm for 10 min, washed with potassium phosphate buffer (pH 6.4) and stored in fresh buffer at 4 °C. Then, the sample was re-assayed for the second cycle after 96 h and the third cycle after 120 h. Similarly, Ca-alginate beads immobilized with β-glucosidase were removed and washed thoroughly with distilled water and stored in fresh buffer at 4 °C, followed by multiple uses for three times as above.

2.10 Desorption Study

Immobilized β-glucosidase–MWCNT biocatalyst was dispersed in 100 mM potassium phosphate buffer (pH 6.4) at room temperature in a shaker at 150 rpm for 6 h. A portion of the dispersed biocatalyst was then removed every hour and separated by centrifugation at 2000 rpm for 20 min. The enzymatic activity of the centrifuged immobilized biocatalyst was measured as mentioned in Sect. 2.7. The difference in activities of the immobilized biocatalyst before dispersion and the one after dispersion at hourly interval amounts to loss of activity over time.

3 Research Findings

3.1 Surface Characterization of MWCNT

The surface morphology of the MWCNT before and after functionalization was observed using scanning electron microscopy (SEM). Figure 2 shows the images of the MWCNT samples observed under ×2000 magnifications. Pristine MWCNT surface is characterized by closely packed MWCNTs with somewhat even surface and uneven interstitial spaces (Fig. 2a). The agglomeration of the MWCNTs indicates the presence of impurities which was remedied by washing with deionized water prior to functionalization. Figure 2b shows the freeze-dried MWCNT sample after functionalization and immobilization of the enzymes. The images show that the interstices and grooves on the surface of the MWCNTs prior to glutaraldehyde

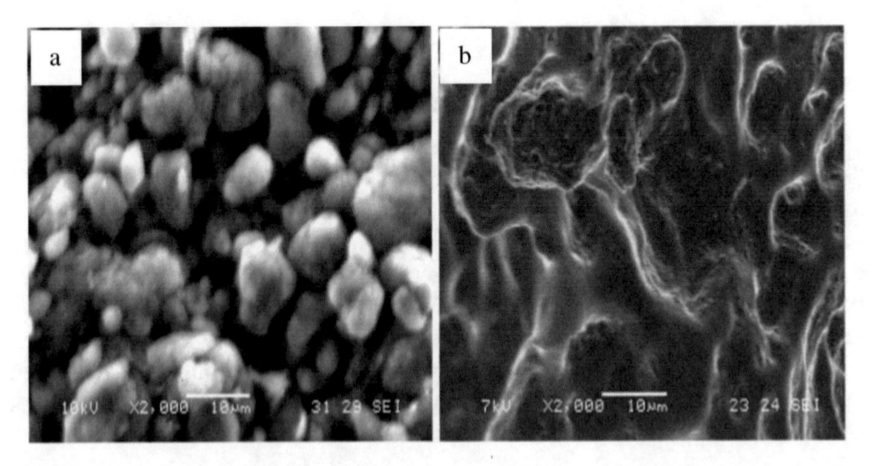

Fig. 2 SEM images of MWCNTs at magnification ×2000: **a** pristine MWCNTs and **b** freeze-dried MWCNTs after glutaraldehyde treatment and β-glucosidase immobilization

modification and β-glucosidase immobilization have been largely filled by enzymes after immobilization [36].

3.2 FTIR Analysis of MWCNT

The FTIR spectra of the MWCNT samples before and after functionalization are shown in Fig. 3. In Fig. 3a, the very low intensity characteristic peaks of the raw MWCNTs were evident in the FTIR spectra at 3447 cm^{-1} (OH) and 1643 cm^{-1} (C=O) [41, 42]. The peak at 1420 cm^{-1} indicates the graphite structure (C=C stretching associated with sidewall defect) in MWCNTs [43]. After functionalization (Fig. 3b), the peaks appear at higher intensity which indicates formation of MWCNT-COOH bond. The peak at 1078 cm^{-1} could be attributed to the = C–H bond. The intensity of the OH and C=O peaks increased after glutaraldehyde treatment indicating higher presence of carboxylic group facilitating the enzyme immobilization [36].

3.3 Enzyme Activity

The effect of immobilization on the activity of β-glucosidase with varying substrate concentration is depicted in Fig. 4. Free enzyme exhibits higher activity compared to immobilized enzyme. The loss of activity in the case of enzyme immobilized on MWCNT is marginal compared to the immobilization in Ca-alginate.

Maximum activity of 19.65 μmol/mL min was exhibited by the free enzyme. Enzyme immobilized in Ca-alginate displayed an activity of 15.06 μmol/mL.min, i.e., 76.64% of the free enzyme activity, while MWCNT-immobilized enzyme retained 19.13 μmol/mL.min activity, i.e., 97.35% of the maximum free enzyme activity (Fig. 5). However, the trend in Fig. 4 clearly shows that most of the enzyme activity in both methods of immobilization is achieved around 2 μmol/mL substrate concentration and no significant change in activity is observed after 3 μmol/mL substrate concentration.

Enzyme immobilized onto MWCNT achieved 17.71 μmol/mL.min activity equivalent to 92.58% of its maximum activity at the substrate concentration of 1 μmol/mL compared to the 13.07 μmol/mL.min equivalent to 86.79% relative activity achieved by enzyme immobilized in Ca-alginate. This can be attributed to the easy access of the enzyme active sites by the substrate molecules due to minimal mass transfer resistance in case of MWCNT support. In the entire range of substrate concentrations studied, the activity of the enzyme immobilized in Ca-alginate appears to be significantly lower compared to that immobilized on MWCNT. This is attributed to the diffusional resistance to substrate transport within the alginate gel matrix having small pore sizes, and hence hindering the enzyme–substrate accessibility [13, 39]. However, a recent study by Tsai and Meyer [44] reported a technique to improve stability of immobilized enzyme in Ca-alginate beads by cross-linking β-glucosidase

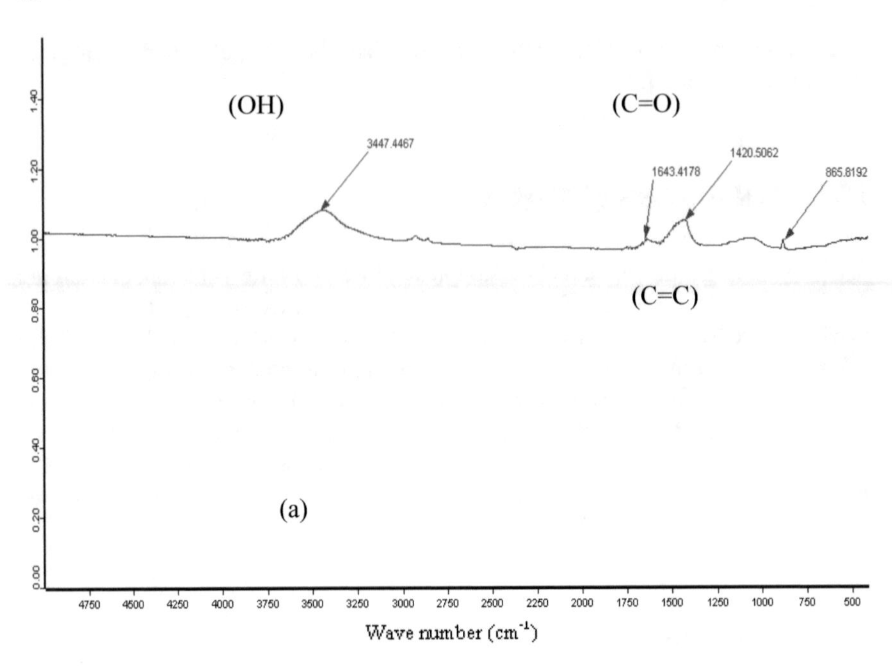

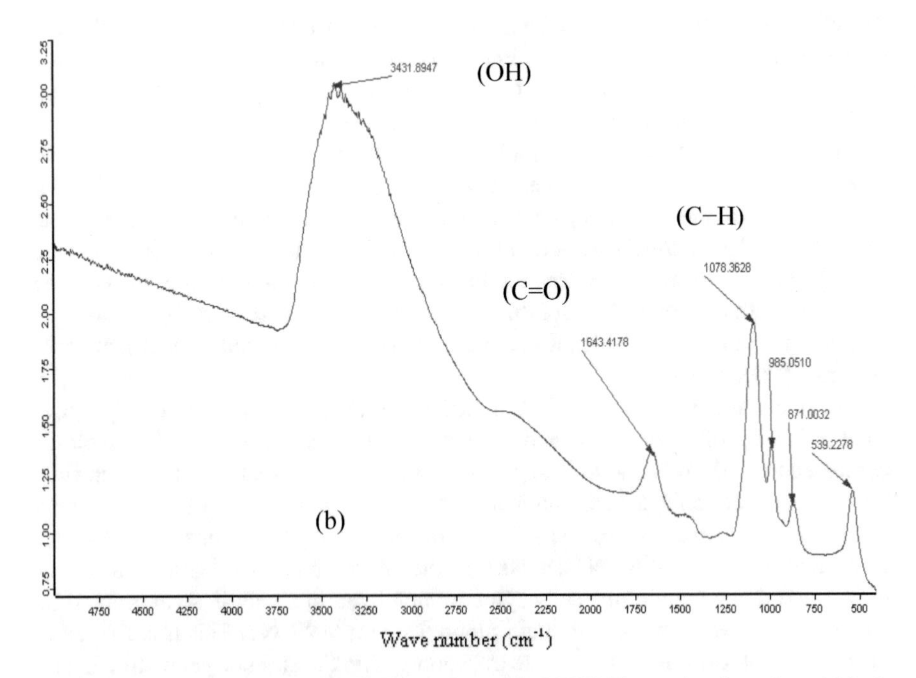

Fig. 3 FTIR spectra: **a** raw MWCNT sample and **b** MWCNT sample after glutaraldehyde functionalization and β-glucosidase immobilization

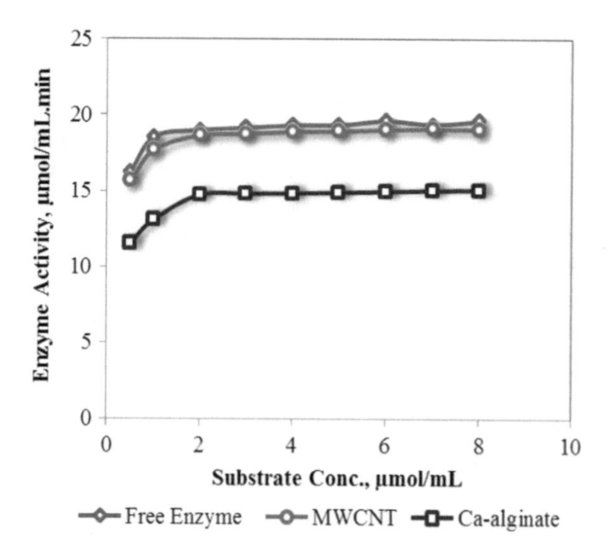

Fig. 4 Dependence of enzyme activity on the substrate concentration for the unbound and the immobilized β-glucosidase with pNPG as the substrate

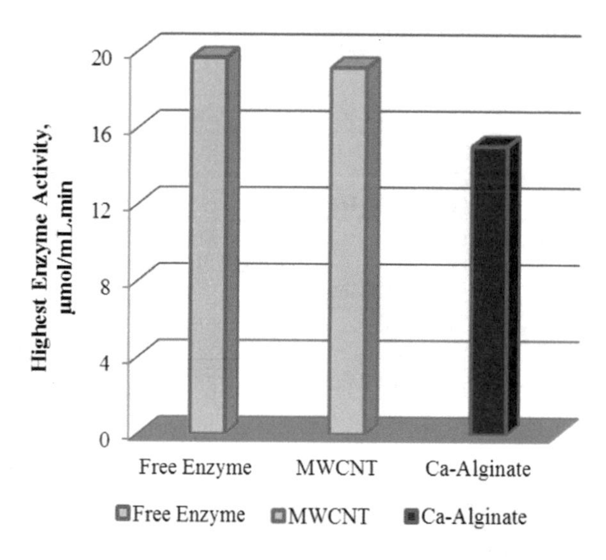

Fig. 5 Comparison of maximum enzyme activities exhibited by the unbound and immobilized β-glucosidase with pNPG as the substrate

with glutaraldehyde prior to entrapment in alginate gel, which showed approximately 60% retention of the enzymatic activity [44].

3.4 Determination of Kinetic Parameters

The plot of enzyme activity against substrate concentration shows typical Michaelis–Menten behavior where saturation occurs at higher substrate concentrations. The

Michaelis–Menten constants K_m and V_{max} were hence determined from Lineweaver–Burk, Hanes (Langmuir) and Eadie–Hofstee linear plots as well as hyperbolic regression plots using Hyper 32 software (Table 1) [45].

With high R^2 values, the Lineweaver–Burk plot provides a better fit for the kinetic data as is evident from Fig. 6, plotted for the case of immobilized MWCNT for the sake of illustration. As expected, the V_{max} obtained for the two immobilized biocatalysts were lower compared to the unbound enzyme. However, immobilization on MWCNT resulted into V_{max} of 19.5 mM/min which is about within 98% of that of the free enzyme. The entrapment of the β-glucosidase within Ca-alginate beads caused significant decrease in V_{max} which was determined to be 15.49 mM/min. On the contrary, the Michaelis constant K_m, for the enzyme, increased after immobilization.

K_m is equal to the substrate concentration required to attain half of the maximum rate of reaction, V_{max}. In other words, K_m is that substrate concentration at which half of the enzyme's active sites are saturated with the substrate. Therefore, K_m is a relative measure of the substrate binding affinity to the enzyme or the stability of

Table 1 Kinetic parameters obtained from various Michaelis-Menten plots

	Lineweaver–Burk	Hanes	Eadie–Hofstee	Hyperbolic regression	Michaelis–Menten
	V_{max} (mM/min) and K_m (mM)				
Free enzyme	19.91 0.1046 ($R^2 = 0.9544$)	19.80 0.0898	19.89 0.1019	19.88 ± 0.2834 0.1005 ± 0.0233	19.80 0.1050
MWCNT	19.50 0.1170 ($R^2 = 0.9846$)	19.35 0.0936	19.49 0.1152	19.48 ± 0.182 0.1137 ± 0.0157	19.49 0.1169
Ca-alginate	15.49 0.1713 ($R^2 = 0.9831$)	15.33 0.139	15.48 0.1698	15.47 ± 0.2397 0.1685 ± 0.0290	15.48 0.1703

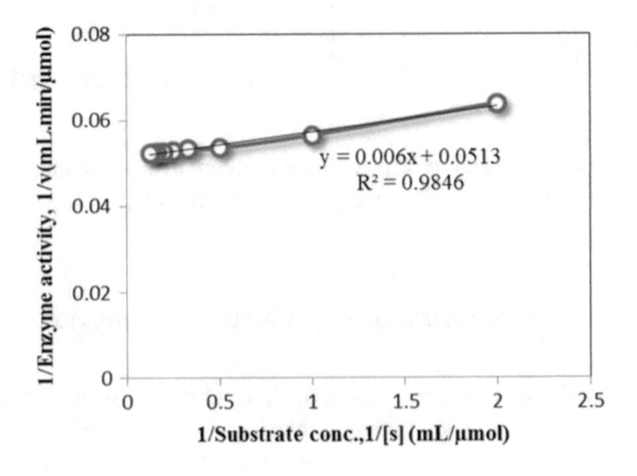

Fig. 6 Lineweaver–Burk plot for β-glucosidase immobilized onto glutaraldehyde-modified MWCNT

the enzyme–substrate complex: Higher K_m values imply lower enzyme affinity for the substrate and vice versa [40]. Thus, increase in K_m value for the Ca-alginate-immobilized enzyme compared to free enzyme amounts to decreased enzyme affinity for the substrate.

3.5 Reusability of Immobilized β-Glucosidase

One of the key factors that determine the effectiveness of the immobilization technique is the reusability of the immobilized enzymes. It also indicates the operational stability and hence economic potential of the immobilized enzyme for industrial applications. As evident in Fig. 7, the enzyme activity is reduced after each assay due to separation, washing and storage management steps. This increased loss in residual activity on subsequent hydrolytic assay is supported by other works as well [13, 34, 43, 46]. The MWCNT immobilization retains 85.6% of its initial activity after third consecutive usage, which is comparable to studies reporting up to 89% retained activity upon 6th repeated usage [34]. Enzymes immobilized in alginate beads lose about 50% of its initial activity by the third cycle [36, 47]. Previous studies conducted on the reusability of enzymes encapsulated by Ca-alginate beads have reported loss in activity upto 35% or higher on second consecutive usage [13, 39, 46]. Jameel et al. [47] have reported 50% activity loss of β-glucosidase immobilized in alginate after the third hydrolysis cycle. This increased loss in activity observed in the beads is caused by the leaching of the enzymes during subsequent assay and washing steps. The MWCNT immobilization provided higher reusability compared to the Ca-alginate beads. In a recent study on the β-glucosidase immobilization on hydroxyapatite nanoparticles, it was possible to retain 70% of the initial activity of the immobilized β-glucosidase after 10 hydrolysis cycles [32]. Verma

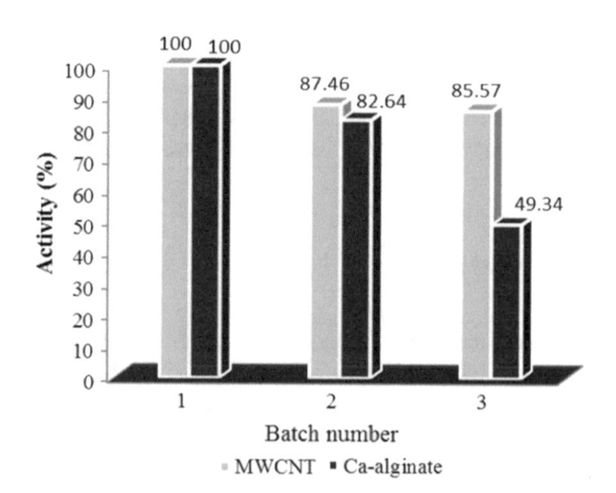

Fig. 7 Comparison of the residual activity after subsequent reuse cycles

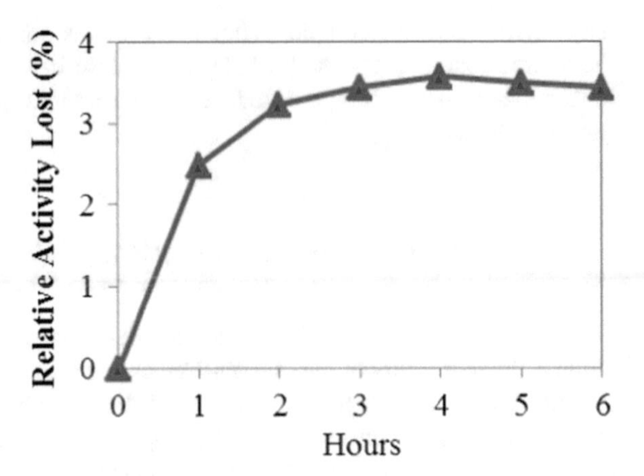

Fig. 8 Desorption of β-glucosidase immobilized on glutaraldehyde-modified MWCNTs

et al. [25] observed residual enzyme activity more than 50% for their immobilized nanoparticle–enzyme conjugate at the 16th cycle reuse.

3.6 Desorption of β-Glucosidase Immobilized on Glutaraldehyde-Modified MWCNT

The desorption study is performed to observe the pattern in enzyme loss from the MWCNT surface over the time. The loss in activity versus time of desorption is shown in Fig. 8. The most activity loss occurs within the first hour of desorption time and becomes almost constant after 2 h [36]. Similar results have been reported in the literature where negligible amount of enzyme was desorbed even after 6 h [34]. In the first hour, some activity loss occurs due to the fact that some of the weakly adsorbed enzymes onto the MWCNT surface are readily desorbed during early desorption process. After the loosely attached enzymes are lost in the beginning, the enzyme activity is retained almost constant over the time due to strong covalent binding between β-glucosidase and functionalized MWCNT. This desorption trend is a testimony to stable binding of enzyme to MWCNT.

4 Conclusion

Performance of immobilized β-glucosidase onto MWCNT and in Ca-alginate beads has been compared. Immobilization on MWCNT has resulted into retention of higher enzyme activity, i.e., 97.35% of the unbound enzyme compared to Ca-alginate-immobilized β-glucosidase exhibiting 76.64%. MWCNT-immobilized enzyme has displayed high V_{max} and lower K_m values of 19.50 mM/min and 0.1170 mM, respectively, in contrast to 15.49 mM/min and 0.1713 mM as corresponding values for

Ca-alginate. This trend is a clear indication of greater accessibility of substrate to enzyme's active sites with lesser diffusional resistance, and high substrate affinity for enzyme in case of MWCNT support in contrast to Ca-alginate. Reusability study delineates more stable immobilization onto MWCNT that could retain 85.57% of its original activity after the third cycle of hydrolysis with corresponding value of 49.34% for Ca-alginate. This substantially high residual activity in multiple uses of MWCNT biocatalyst compared to the enzyme–Ca-alginate conjugate is indeed significant in overall process economy. Despite various fascinating advantages that MWCNT as enzyme support has to offer, there still remain challenges for efficient recovery of the biocatalyst in the reactor and optimization of the immobilization process, which demands more concerted research. However, this study provides distinct economic advantage of employing nanoparticles as support for enzyme immobilization for large-scale industrial applications.

Acknowledgements The research endowment fund grant No. EDW B13-036-0921 by the International Islamic University Malaysia is gratefully acknowledged. The authors thank Elsevier Inc. for granting permission to reproduce figure from the book *Methods in Enzymology* (Elsevier Inc.) as Fig. 1 in the present manuscript.

Conflict of Interest The authors of this chapter have no conflict of interest.

References

1. W.E. Workman, D.F. Day, Purification and properties of β-glucosidase from *Aspergillus Terreus*. Appl. Environ. Microbiol. **44**(6), 1289–1295 (1982)
2. A. Sørensen, P.S. Lübeck, M. Lübeck, P.J. Teller, B.K. Ahring, β-Glucosidases from a new *Aspergillus* Species can substitute commercial β-glucosidases for saccharification of lignocellulosic biomass. Can. J. Microbiol. **57**(8), 638–650 (2011)
3. A. Martino, P.G. Pifferi, G. Spagna, The separation of pectinlyase from β-glucosidase in a commercial preparation. J. Chem. Technol. Biotechnol. **61**(3), 255–260 (2013)
4. M.C. Moscatelli, β-Glucosidase kinetic parameters as indicators of soil quality under conventional and organic cropping systems applying two analytical approaches. Ecol. Ind. **13**(1), 322–327 (2012)
5. A. Bai, X. Zhao, Y. Jin, G. Yang, Y. Feng, A novel thermophilic β-glucosidase from *Caldicellulosiruptor Bescii*: Characterization and its synergistic catalysis with other cellulases. J. Mol. Catal. B Enzym. **85–86**, 248–256 (2013). https://doi.org/10.1016/j.molcatb.2012.09.016
6. Y. Lin, S. Tanaka, Ethanol fermentation from biomass resources: current state and prospects. Appl. Microbiol. Biotechnol. **69**, 627–642 (2006)
7. A. Celik, A. Dincer, T. Aydemir, Characterization of β-glucosidase immobilized on chitosan-multiwalled carbon nanotubes (MWCNTs) and their application on tea extracts for aroma enhancement. Int. J. Biol. Macromol. **89**, 406–414 (2016)
8. Q. Wang, L. Zhou, Y. Jiang, J. Gao, Improved stability of the carbon nanotubes-enzyme bioconjugates by biomimetic silicification. Enzyme Microbial Technol. **49**(1), 11–16 (2011)
9. R.E. Abraham, M. Puri, Nano-immobilized cellulases for biomass processing with application in biofuel production, in *Methods in Enzymology* (Elsevier Inc., 2019). https://doi.org/10.1016/bs.mie.2019.09.006
10. R. Ahmad, S.K. Khare, Immobilization of *Aspergillus niger* cellulase on multiwall carbon nanotubes for cellulose hydrolysis. Biores. Technol. **252**, 72–75 (2018). https://doi.org/10.1016/j.biortech.2017.12.082

11. M. Khan, Q. Husain, Multiwalled carbon nanotubes bound beta-galactosidase: It's activity, stability and reusability. Methods Enzymol. (2019). https://doi.org/10.1016/bs.mie.2019.10.018
12. N. Ortega, M.D. Busto, M. Perez-Mateos, Optimization of β-glucosidase entrapment in alginate and polysaccharide. Biosource Technol. **64**, 105–111 (1998)
13. A. Anwar, S. Ali Ul Qader, S. Iqbal, A. Riaz, A. Azhar, Calcium alginate: a support material for immobilization of proteases from newly isolated strain of *Bacillus Subtilis* KIBGE-HAS. World Appl. Sci. J. 7(10), 1281–1286 (2009)
14. D.B. Magalhaes, M.H. Miguez da Rocha-Leao, Immobilization of β-glucosidase aggregates in calcium alginate. Biomass Bioenerg. **1**(4), 213–216 (1991)
15. M.D. Busto, N. Ortega, M. Perez-Mateos, Studies on microbial β-D-glucosidase immobilized in alginate gel beads. Process Biochem. **30**(5), 421–426 (1995)
16. S. Erzheng, X. Tao, G. Liping, D. Qianying, Z. Zhengzhu, Immobilization of β-glucosidase and its aroma-increasing effect on tea beverage. Food Bioprod. Process. **88**, 83–89 (2010)
17. P. Zheng, J. Wang, C. Lu, Y. Xu, Z. Sun, Immobilized β-glucosidase on magnetic chitosan microspheres for hydrolysis of straw cellulose. Process Biochem. **48**(4), 683–687 (2013)
18. I.S. Tan, K.T. Lee, Immobilization of β-glucosidase from *Aspergillus niger* on κ-carrageenan hybrid matrix and its application on the production of reducing sugar from macroalgae cellulosic residue. Biores. Technol. **184**, 386–394 (2015). https://doi.org/10.1016/j.biortech.2014.10.146
19. J. Zhang, D. Wang, J. Pan, J. Wang, H. Zhao, Q. Li, X. Zhou, Efficient resveratrol production by immobilized β-glucosidase on cross-linked chitosan microsphere modified by L-lysine. J. Mol. Catal. B Enzym. **104**, 29–34 (2014)
20. T.M. Silva, B.C. Pessela, J.C.R. Silva, M.S. Lima, J.A. Jorge, J.M. Guisan, M.L.T.M. Polizeli, Immobilization and high stability of an extracellular β-glucosidase from *Aspergillus japonicus* by ionic interactions. J. Mol. Catal. B Enzym. **104**, 95–100 (2014). https://doi.org/10.1016/j.molcatb.2014.02.018
21. L.T. Nguyen, Y.S. Lau, K.-L. Yang, Entrapment of cross-linked cellulase colloids in alginate beads for hydrolysis of cellulose. Colloids Surf. B **145**, 862–869 (2016)
22. Z. Ding, X. Zheng, S. Li, X. Cao, Immobilization of cellulase onto a recyclable thermo-responsive polymer as bioconjugate. J. Mol. Catal. B Enzym. **128**, 39–45 (2016). https://doi.org/10.1016/j.molcatb.2016.03.007
23. W. Feng, Enzyme immobilized on carbon nanotubes. Biotechnol. Adv. **29**, 889–895 (2011)
24. M.L. Verma, M. Naebe, C.J. Barrow, M. Puri, Enzyme immobilization on amino-functionalized multi-walled carbon nanotubes: structural and biocatalytic characterization. PLoS One 1371–1382 (2013a)
25. M.L. Verma, R. Chaudhary, T. Tsuzuki, C.J. Barrow, M. Puri, Immobilization of β-glucosidase on a magnetic nanoparticle improves thermostability: application in cellobiose hydrolysis. Biores. Technol. **135**, 2–6 (2013)
26. S.A. Ansari, Q. Husain, Potential application of enzymes immobilized on/in nano materials: a review. Biotechnol. Adv. **30**, 512–523 (2012)
27. M. Patila, N. Chalmpes, E. Dounousi, H. Stamatis, D. Gournis, Use of functionalized carbon nanotubes for the development of robust nanobiocatalysts. Methods Enzymol. (2019). https://doi.org/10.1016/bs.mie.2019.10.015
28. P. Kumar, B. Ryan, G.T.M. Henehan, β-Glucosidase from *Streptomyces griseus*: nanoparticle immobilization and alkyl glucoside synthesis. Protein Expr. Purif. **132**, 164–170 (2017)
29. K. Khoshnevisan, F. Vakhshiteh, M. Barkhi, H. Baharifar, E. Poor-Akbar, N. Zari, A.-K. Bordbar, Immobilization of cellulase enzyme onto magnetic nanoparticles: applications and recent advances. Mol. Catal. **442**, 66–73 (2017). https://doi.org/10.1016/j.mcat.2017.09.006
30. O.M. Darwesh, S.S. Ali, I.A. Matter, T. Elsamahy, Y.A. Mahmoud, Enzymes immobilization onto magnetic nanoparticles to improve industrial and environmental applications. Methods Enzymol. (2019). https://doi.org/10.1016/bs.mie.2019.11.006. (Elsevier)
31. K. Selvam, M. Govarthanan, D. Senbagam, S. Kamala-Kannan, B. Senthilkumar, T. Selvankumar, Activity and stability of bacterial cellulase immobilized on magnetic nanoparticles. Chin. J. Catal. **37**(11), 1891–1898 (2016). https://doi.org/10.1016/s1872-2067(16)62487-7

32. T.C. Coutinho, M.J. Rojas, P.W. Tardioli, E.C. Paris, C.S. Farinas, Nanoimmobilization of β-glucosidase onto hydroxyapatite. Int. J. Biol. Macromol. **119**, 1042–1051 (2018)
33. N.S. Valentina, β-Galactosidase entrapment in silica gel matrices for a more effective treatment of lactose intolerance. J. Mol. Catal. B Enzym. **71**, 10–15 (2011)
34. S.A. Ansari, R. Sattar, S. Chibber, M.J. Khan, Enhanced stability of kluyveromyces lactis beta-galactosidase immobilized on glutaraldehyde modified multiwalled carbon nanotubes. J. Mol. Catal. B Enzym. **97**, 258–263 (2013)
35. I. Migneault, C. Dartiguenave, M.J. Bertrand, K.C. Waldron, Glutaraldehyde behavior in aqueous solution, reaction with proteins, and application to enzyme crosslinking. Biotechniques **37**, 790–802 (2004)
36. L. Mahmood, Comparative study of the effectiveness of immobilized β-glucosidase enzyme on CNT-nanoparticles and Ca-alginate beads. Undergraduate Dissertation, Biotechnology Engineering Department, International Islamic University Malaysia, Kuala Lumpur, 2014
37. C. Mateo, J.M. Palomo, G. Fernandez-Lorente, J.M. Guisan, R. Fernandez-Lafuente, Improvement of enzyme activity, stability and selectivity via immobilization techniques. Enzyme Microbial Technol. **40**, 1451–1463 (2007)
38. Y.R. Jung, H.Y. Shin, Y.S. Song, S.B. Kim, S.W. Kim, Enhancement of immobilized enzyme activity by pretreatment of β-glucosidase with cellobiose and glucose. J. Ind. Eng. Chem. **18**, 702–706 (2012)
39. A.T. Jameel, Y. Faridah, S. Johana, Performance of β-glucosidase immobilized on calcium alginate beads, in *Proceedings of the International Conference on Biotechnology Engineering, ICBioE 2013* (International Islamic University Malaysia, Kuala Lumpur, July 2013), pp. 655–661
40. P.M. Doran, *Bioprocess Engineering Principles*, 2nd edn. (Academic Press, Elsevier Ltd., Oxford UK, 2013), pp. 661–623
41. N.G. Sahoo, H. Bao, Y. Pan, M. Pal, M. Kakran, H.K.F. Cheng, L. Li, L.P. Tan, Functionalized carbon nanomaterials as nanocarriers for loading and delivery of poorly water soluble anticancer drug: A comparative study. Chem. Commun. **47**, 5235–5237 (2011). https://doi.org/10.1039/c1cc00075f
42. L.Z.W. Qi, Improved stability of the carbon nanotubes-bioconjugates by biomimetic silification. Enzyme Microbial Technol. **49**, 11–16 (2011)
43. N.M. Mubarak, J.R. Wong, K.W. Tan, J.N. Sahu, E.C. Abdullah, N.S. Jayakumar, P. Ganesan, Immobilization of cellulase enzyme on functionalized multiwall carbon nanotubes. J. Mol. Catal. B Enzym. **107**, 124–131 (2014)
44. C.-T. Tsai, A.S. Meyer, Enzymatic cellulose hydrolysis: enzyme reusability and visualization of β-glucosidase immobilized in calcium alginate. Molecules **19**(12), 19390–19406 (2014)
45. *Hyper 32 Software* (University of Liverpool, 2012)
46. S.A. Qader, Characterization of dextransucrase immobilized on calcium alginate beads from leuconostoc mesenteroides PCSIR-4. Ital. J. Biochem. **56**, 158–162 (2007)
47. A.T. Jameel, K.Y. Maalim, F. Yusof, Relative characterization of the immobilized beta-glucosidase on Ca-alginate and acid functionalized-multiwalled carbon nanotubes. Jurnal Teknologi (Sci. Eng.) **81**(3), 1–10 (2019). https://doi.org/10.11113/jt.v81.13314

Graphene Synthesis and Antibody Immobilization Techniques for Immunosensors

Ihda Uswatun Shalihah Shohibuddin, Piravin Raj Barthasarathy, and Wan Wardatul Amani Wan Salim

Abstract The principle of antigen–antibody interaction is exploited for immunosensor development. Over the years, immunosensors have been fabricated for different applications, and the fabrication process has benefited from the use of nanomaterials. Graphene, a single-atom-thick layer of carbon, has been used to coat the sensing electrodes of immunosensors; antibodies are then immobilized onto the graphene-modified electrodes to produce graphene-based immunosensors. Here, we describe several techniques for producing graphene and its derivatives. We also focus on approaches for antibody immobilization on these graphene-modified electrodes.

Keywords Graphene · Immunosensor · Antibody immobilization · Graphene-modified electrodes

1 Introduction

Graphene is an allotrope of carbon in which the carbon atoms are arranged hexagonally in a 2D honeycomb crystal lattice that is one-atom-thick [1]. The 2D structure can be chemically modified and made to wrap into fullerene, rolled into nanotube, or stacked in several layers [2]. Graphene is a zero-band gap material; the band gap determines the minimum energy required to excite an electron from a valence band to a conduction band. When electrons make such a band jump, they possess enough energy to move freely and participate in conduction. Since both bands in graphene touch at six points called the Dirac points, the electrons can move freely between these bands [3]. Graphene shows electron mobility with values in excess of 200,000 $cm^2\ V^{-1}\ s^{-1}$ at electron densities of $\sim 2 \times 10^{11}\ cm^{-2}$, indicating good electrical conductivity [4]. However, graphene oxide (GO) flakes have a band gap of 0.7 eV [5] that decreases with reduction of the oxygen/carbon (O/C) ratio. Reduced graphene oxide (rGO) has a band gap of 3.39 eV for a 50% O/C ratio; however, it

I. U. S. Shohibuddin · P. R. Barthasarathy · W. W. A. Wan Salim (✉)
Department of Biotechnology Engineering, Faculty of Engineering, International Islamic University Malaysia, 50728 Gombak, Kuala Lumpur, Malaysia
e-mail: asalim@iium.edu.my

A. T. Jameel and A. Z. Yaser (eds.), *Advances in Nanotechnology and Its Applications*,
https://doi.org/10.1007/978-981-15-4742-3_2

decreases to 1.44 eV with 25% O/C ratio [6]. This indicates rGO is more conductive than GO flakes.

2 Graphene Synthesis Methods

In 2004, Andre Geim and Constantine Novoselov earned a Nobel Prize for isolating graphene from graphite using adhesive tapes [1]. Here, we describe three methods on graphene synthesis: mechanical exfoliation, chemical vapor deposition (CVD), and Hummer's method.

2.1 Mechanical Exfoliation

Graphite is composed of several layers of graphene that are bound together by weak van der Waals forces [7]; hence, mechanical force of ~300 nN/μm^2 can be applied to exfoliate a single layer of graphene from graphite [8]. Using adhesive tape to peel graphene from graphite requires repeated peeling, and the number of peelings to obtain a single layer of graphene is difficult to estimate [1]. Huc et al. reported a reverse mechanical exfoliation technique to obtain graphene from highly oriented pyrolytic graphite (HOPG). The method of graphene synthesis through reverse mechanical exfoliation is described [9] below.

Materials
Freshly cleaved bulk HOPG; silicon chip; and epoxy glue, scalpel, and scotch tape.

Methods

1. Prepare HOPG by pre-cutting the bulk HOPG to a thickness of <100 μm.
2. Cover the top layer of oxidized silicon chip with epoxy glue.
3. Place HOPG on the epoxy glue. Compress the HOPG on the silicon chip using a screw-press technique.
4. Cleave the bulk HOPG with the scalpel.
5. Attach scotch tape to HOPG, and peel it off; the last layer of graphene will be strongly attached to the epoxy. The graphene sheet obtained is the first side that adhered to the epoxy after the curing step.

The Raman spectroscopy showed a strong band at 1582 cm^{-1} which indicates multiple layers of graphene sheets. Meanwhile, smaller intensity band at 1587 cm^{-1} indicates a monolayer structure. Therefore, the reverse exfoliation method has proven to produce flat and large graphene flakes due to the strong bond between the graphite underlayer to the epoxy-based glue substrate. Tens of microns wide of single-layer graphene sheets were successfully obtained; this technique opens a way toward graphene deposition on any substrates.

2.2 Chemical Vapor Deposition (CVD)

CVD is a technique used to manufacture graphene of uniform thickness. Features of the graphene layer can be modified by adjusting the temperature and duration of the process. The graphene layer is produced through deposition of gaseous reactants onto a heated substrate within the CVD chamber [10].

Faggio et al. investigated graphene production using ethanol-assisted CVD technique. The graphene was produced by varying three experimental parameters. The parameters studied are growth temperature (1000 and 1070 °C), reaction time (10 and 30 min), and hydrogen flow rate ($\Phi_{H_2} = 0$, 1, 10, and 100 sccm). The production of graphene is divided into Stage 1 and Stage 2, in which ethanol was supplied to CVD system during Stage 2. Argon gas acts as a carrier for ethanol flow into the CVD system. This is possible as the ethanol that is kept in a steel bubbler vessel is pressurized in argon at 3 bar. The steel bubbler vessel is stored at 0 °C, with the ethanol at equilibrium pressure of 0.015 bar. The amount of ethanol entering the CVD system is calculated as follows: (0.015/3 bar) $\times$ 100% = 0.5%. During Stage 2, the flow rate of argon is set at 20 sccm. Hence, the ethanol enters the vessel at a flow rate of 0.1 sccm. The method on graphene synthesis through CVD is described [11] below.

Materials

Copper foil (99.95% purity); pure argon (Ar) and hydrogen (H_2) gases; ethanol that is kept in a steel bubbler vessel; and CVD system containing quartz tube (as a reaction chamber).

Method

Stage 1:

1. Cut copper foil to a thickness of 25 μm.
2. Heat up the CVD system consisting of a high-vacuum fitted tube furnace.
3. Allow the flow of 20 sccm of Ar and 20 sccm of H_2 into the CVD chamber.
4. Place the copper foil on the 2-m-long quartz tube, and place the tube in the hot zone of the CVD chamber.
5. Allow the sample to anneal for 20 min at high temperature (1000/1070 °C).

Stage 2:

1. Set desired flow rate of H_2 ($\Phi_{H_2} = 0$–100 sccm) into the CVD system.
2. Switch to Ar flow from the steel bubbler vessel that contains ethanol, at a flow rate of 20 sccm.
3. Allow the gaseous reactant to deposit on the substrate for 10–30 min.
4. Allow the sample to cool down before extraction of carbon film. Clean the carbon film using a stationary rubber eraser, optics cleaning tissues, and ethanol. Allow the carbon film to float for 2 h at room temperature.
5. Scoop up the carbon film using a thermally oxidized silicon wafer.

The results suggest that hydrogen flow rate and temperature have strong effects on the graphene growth. The optimal parameters of 10 sccm hydrogen flow rate, 1070 °C high temperature, and 10 min short growth time are required to synthesize a monolayer or bilayer crystallized graphene film. Importantly, ethanol can be a substitute to other hydrocarbons such as methane in the CVD graphene synthesis.

2.3 *Chemical Method (Hummer's Method)*

William S. Hummers and Richard E. Offeman pioneered the production of graphite oxide through Hummer's method [12], which was patented in 1957 [13]. A combination of sulfuric acid (H_2SO_4), potassium permanganate ($KMnO_4$), and sodium nitrate ($NaNO_3$) is used to produce GO from graphite. The synthesis temperature is below 100 °C, which reduces production cost. However, Hummer's method produces toxic gases such as nitrogen dioxide (NO_2) and dinitrogen tetroxide (N_2O_4). In addition, it is also difficult to remove Na^+ and NO_3^- ions from the wastewater produced during GO synthesis and purification stage. Therefore, an eco-friendlier approach eliminating the use of sodium nitrate for GO production was proposed. The modification does not significantly affect the yield of GO; most importantly, the method is suitable for GO production in large quantity [14].

Choudhary et al. used CVD techniques to produce GO to coat sensor electrodes. The sensor was able to detect the expression of human telomerase reverse transcriptase. The method for graphene synthesis by Hummer's method is described [15] below.

Materials
Graphite powder; sodium nitrate ($NaNO_3$) and sulfuric acid (H_2SO_4); potassium permanganate ($KMnO_4$) flakes; hydrogen peroxide (H_2O_2); and Millipore water, distilled water, and ice water bath.

Method

1. Dissolve 1 g graphite powder and 1 g $NaNO_3$ in 46 mL H_2SO_4 with continuous stirring in an ice bath.
2. Add 6 g $KMnO_4$ flakes while still stirring.
3. Once the solution turns dark green, transfer to a water bath (35 ± 5 °C). Continue to stir the solution vigorously.
4. Raise the temperature to 90 ± 5 °C. Continue to stir the solution while adding 96 mL Millipore water.
5. Add 200 mL distilled water.
6. Slowly add 6 mL of H_2O_2.
7. Filter the solution.
8. Wash the solution with 200 mL distilled water.
9. Agitate the filter cake to disperse in water mechanically.

10. Centrifuge at 200 rpm for 2 min.
11. Centrifuge again at 800 rpm for 15 min.
12. Sonicate the final sediment to disperse in water.

Atomic force microscope (AFM) image revealed a homogeneous distribution of GO film on silicon wafer; lateral dimension of the GO film was ~2500 nm for the larger sheets and ~200 nm for the smaller sheets. The thickness of a monolayer of GO sheet was ~0.87 nm.

3 Immunosensors

Antibody binds to antigens through weak chemical interaction contributed by electrostatic forces, hydrophobic, van der Waals, and ionic interactions, forming an antigen–antibody complex. The antigen–antibody interaction is analogous to a lock and key model, whereby the antigen binding site, or paratope, of the antibody complements the antibody binding site, or epitope, of the antigen. Since a weak non-covalent bond holds the antigen and antibody together, the interaction is reversible. In the fabrication of an immunosensor, the antibody is immobilized on the immunosensor for antigen detection. The type of antibody used is dependent on the analyte to be tested.

This section introduces simple antibody immobilization strategies for the active sensing electrodes of a biosensor for immunosensing. The active sensing electrode is the working electrode in a three-electrode electrochemical cell. The three-electrode system consists of working, counter, and reference electrodes. Before antibody immobilization, the working electrode of a biosensor are modified with nanomaterials such as graphene to enhance electron transfer at electrode–solution interface.

3.1 Graphene-Based Immunosensors

Graphene and its derivatives are used to coat the sensing or working electrodes of a three-electrode system in an electrochemical cell. GO and rGO exhibit remarkable physicochemical properties, which are highly preferred for sensor fabrication. These materials are often added along with other materials and tuned for desired sensing performance (e.g., low detection limit).

Physicochemical properties can be defined as physical properties (anything that can be observed and measured, in which the property describes the state of physical system), solvation properties (interaction between solute and solvent molecules), and chemical properties that govern the change in chemical composition as a material takes part in a chemical reaction. The working electrode can be modified with a coating of graphene, GO, rGO, and/or their nanocomposites with the intention to improve electron transfer at the electrode–solution interface. Immunosensor devices with such modified electrodes can potentially offer better sensing performance in terms of lower detection limit.

Several immunosensors have been developed to detect and quantify *Escherichia coli* O157:H7. Wang et al. have developed impedimetric immunosensor with a gold nanoparticle-modified free-standing reduced graphene oxide paper sensing electrode in a three-electrode system. The sensor reported a detection limit of 1.5×10^2 CFU/mL [16]. Mo et al. modified working electrode of screen-printed carbon electrode (SPCE) with reduced graphene oxide-neutral red-gold/platinum nanoparticles. The detection through cyclic voltammetry method reported a detection limit of 2.84×10^3 CFU/mL [17]. Chang et al. developed field-effect transistor device with a sensing electrode that has been modified with thermally reduced monolayer graphene oxide sheets. The immunosensor reported a detection limit of 10 CFU/mL [18].

3.2 Antibody Immobilization Strategies

Immunosensor fabrication involves antibody immobilization onto a sensing or working electrode. During immobilization, immunological activity of the antibody must be retained. The immobilization step aims to orient antibodies so that the antigen binding sites are exposed to the analyte; such antibody orientation can influence the detection limit of an immunosensor. Antibody orientation has been studied using atomic force microscopy [19], time-of-flight secondary-ion mass spectrometry [20], and neutron reflectometry [21]. The heavy-chain F_c region of an antibody needs to be attached to the electrode surface, whereas the light-chain F_{ab} region should be freely available to interact with antigens [22]. Here, we describe three antibody immobilization strategies: physical adsorption, covalent attachment, and affinity interactions.

3.2.1 Physical Adsorption

Physical adsorption refers to direct adsorption of antibodies onto the immunoassay sensing or working electrode, as shown in Fig. 1. Typically, nanomaterials are deposited on an electrode and antibodies are drop-cast onto the modified electrodes, followed by an incubation step, after which loosely bound antibodies are rinsed off with distilled water. The interaction between antibodies and nanomaterial-modified electrodes is driven by weak electrostatic interaction and hydrophobic bonds.

This method, however, has two major limitations: The weak interaction allows antibodies to leach out owing to changes in parameters such as pH and temperature, and antibodies are oriented in a random manner, compromising the sensitivity and selectivity of immunosensors. Hence, physical adsorption is not the method of choice when a low detection limit is required [23].

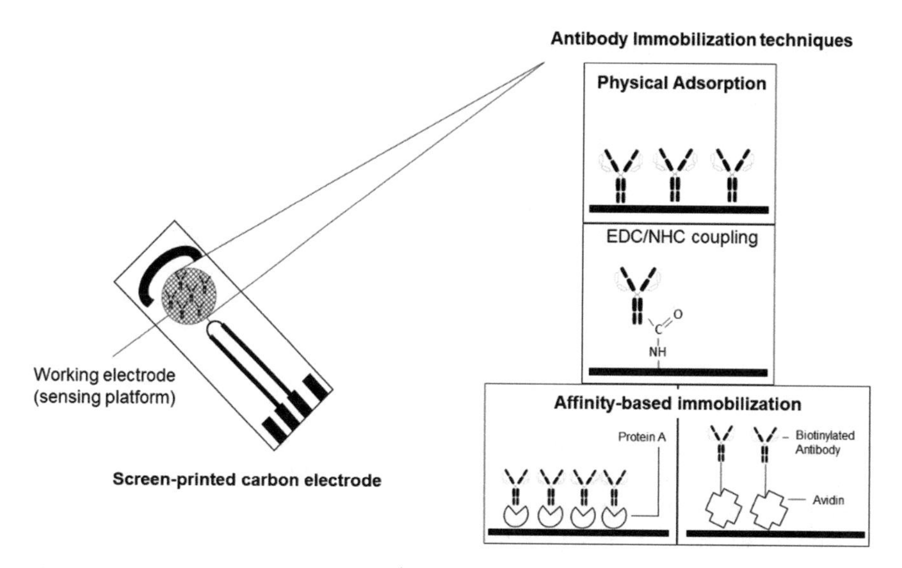

Fig. 1 Screen-printed carbon electrode for immunosensing; the antibodies are immobilized on the working electrode through immobilization techniques of physical adsorption, EDC/NHS coupling method, protein A-mediated immobilization, or avidin–streptavidin. For illustration purposes, all antibodies are oriented in the same direction

3.2.2 Covalent Attachment

Covalent binding can prevent antibody leaching from the electrode surface. Unlike physical adsorption, this method requires chemical reaction between antibodies and the sensing electrodes [24]. Two common covalent immobilization methods will be discussed: carbodiimide-assisted amidation and glutaraldehyde cross-linking.

Carbodiimide-Assisted Amidation

Carbodiimides are cross-linkers that form amide linkages between amines and carboxylates. Specifically, carbodiimide acts as connecting point between electrode surface and antibody. One example of a water-soluble carbodiimide is 1-ethyl-3-[3-dimethylaminopropyl] carbodiimide (EDC). EDC is used together with *N*-hydroxysuccinimide (NHS) or sulfo-NHS (*N*-hydroxysulfosuccinimide) (Fig. 1). When EDC reacts with the carboxyl group of a graphene derivative, the reaction produces o-acylisourea ester. The ester, being unstable in an aqueous medium, immediately reacts with free amide group present on the Fc region of the antibody. Water molecules eventually disrupt the ester and regenerate the carboxyl group. The problem is counteracted by providing NHS or sulfo-NHS that can produce a semi-stable amine-reactive ester [25].

Jung et al. reported on the fabrication of an immunosensor to detect and quantify rotavirus. The rotavirus antibody was immobilized onto a GO-modified sensing electrode using the EDC-NHS coupling method. This method of antibody immobilization through carbodiimide-assisted amidation reaction is described [26] below.

Materials

Amino-modified glass purchased from Nuricell Inc. (Seoul, South Korea); GO solution (0.3 mg/mL); 1-ethyl-3-[3-dimethylaminopropyl] carbodiimide (EDC; 0.5 mM) and *N*-hydroxysulfosuccinimide (sulfo-NHS; 1.0 mM); and rotavirus monoclonal antibody (1 μg/mL).

Methods

1. Drop 1 μL GO solution onto amino-modified glass, and dry the glass surface in a humid chamber.
2. Wash the modified glass with distilled water.
3. Add 0.5 μL EDC and 0.5 μL sulfo-NHS to GO-deposited glass, and incubate at 37 °C for 30 min.
4. Add 1 μL of antibody, and incubate for another 3 h.

AFM images showed the height of rotavirus antibody-linked GO array is 11.9 nm; the theoretical range of an antibody is about 10–15 nm. When rotavirus interacted with the antibody, the height increased from 11.9 to 81 nm. The change in the height confirmed successful interaction of rotavirus to the immobilized antibody as the rotavirus size ranges between 70 and 100 nm. The fabricated immunosensor was tested with different rotavirus concentrations ranging from 10^3 to 10^5 pfu/mL. The limit of detection of the sensor is 10^5 pfu/mL.

Cross-Linking

Cross-linking is a covalent attachment technique that can provide stability to the antibody immobilization. Stability refers to the ability of the antibody, a protein molecule, to be rigid and eventually able to resist conformational changes [27]. Glutaraldehyde (GA) is a commonly used cross-linker that can react with primary amines on antibodies. Graphene-based sensing electrodes need to be modified with free amine groups before glutaraldehyde cross-linking. For biosensing purposes, graphene can be functionalized by introducing bovine serum albumin (BSA) with GA; BSA contains free amine groups, and GA cross-links with the free amine group with cholesterol oxidase and cholesterol esterase [28].

Lin et al. used GA to immobilize anti-carcinoembryonic antigen–antibody on a GO-modified electrode. The immunosensor fabrication procedure used two different solutions. Washing solution was used to remove unbound antibody from the electrode surface. Blocking solution was used to block remaining binding sites on the electrode.

The method for antibody immobilization through glutaraldehyde cross-linking is described [29] below.

Materials
Glassy carbon electrode (GCE) (3-mm diameter); 1.0, 0.3, and 0.05 μm alumina slurry (Buehler); GO solution (0.5 mg/mL); chitosan (0.05% in ultrapure water); glutaraldehyde (2.5%); phosphate-buffered saline, pH 7.4 and pH 8; ultrapure water, deionized water, nitric acid, and acetone; anti-carcinoembryonic antigen (CEA) antibody (0.2 mg/mL); washing solution (Tris-HNO_3 buffer (0.1 M, pH 7.4) containing 0.05% (w/v) Tween 20); and blocking solution (Tris-HNO_3 (0.1 M) containing 5% or 10% (w/v) BSA).

Method

1. Polish GCE using 1.0, 0.3, and 0.05 μm alumina slurry, and wash the electrode with deionized water.
2. Sonicate the electrode successively in nitric acid, acetone, and deionized water.
3. Rinse the electrode with ultrapure water, and allow to dry at room temperature.
4. Drop 5 μL GO solution onto polished GCE, and allow to dry in ambient air.
5. Drop 3 μL of chitosan solution onto the GO-modified GCE, and allow to dry in ambient air.
6. Reduce the GO film electrochemically in PBS, pH 8, at a potential of −1.0 V.
7. Wash the modified electrode with ultrapure water.
8. Incubate the electrode for 2 h with 5 μL glutaraldehyde (in 50 mM PBS, pH 7.4), and wash with ultrapure water.
9. Add 5 μL anti-CEA antibody to the modified electrode surface, and incubate for 1 h at room temperature.
10. Continue incubation overnight in a moisture-saturated environment at 4 °C.
11. Wash the electrode surface with washing solution.
12. Drop 5 μL blocking solution onto the electrode, and incubate electrodes for one hour at room temperature.
13. Wash the electrode again with washing solution.

The limit of detection of the sensor is 0.12 pg/mL. Intra-assay and inter-assay precisions were analyzed using 50 pg/mL of CEA, in which the experiment was repeated for 5 times. A relative standard deviation of 4.8 and 6.5% was obtained for intra-assay and inter-assay, respectively. The sensor was kept in a dry condition at 4 °C. After 2 weeks storage, the sensor was able to retain 90% of the initial response, indicating its stability. The developed immunosensor was able to detect the carcinoembryonic antigen in the serum sample up to picogram concentration, while the commercial electrochemiluminescence test can only detect up to nanogram concentration.

3.2.3 Affinity-Based Immobilization

Affinity-based immobilization controls orientation of antibody, thus providing a high-density F_{ab} region for antigen binding. Examples of affinity-based immobilization are biotin–avidin binding or using immunoglobulin binding bacterial proteins. The biotin–avidin approach is a popular method for immobilizing antibodies onto sensing electrodes. Avidin is a glycoprotein consisting of four identical subunits, in which each unit can bind to biotin [30]. Streptavidin is an analog of avidin, with the ability to bind to four biotins [31]. Electrode surfaces are functionalized with avidin molecules; biotinylated antibodies bind to the avidin-treated electrode surface (Fig. 1) [32].

Immunoglobulin binding bacterial protein is used to immobilize antibodies onto a sensing electrode. The protein binds to the F_c region of an antibody. The two commonly used immunoglobulin binding bacterial proteins are proteins A and G [33], which bind to the antibody strongly at pH 8 [34] and pH 4 to 5 [35], respectively.

Wang et al. reported an impedimetric immunosensor for detection of *E. coli* O157:H7. The researchers used wash buffer to remove unbound and excessive streptavidin molecules from the electrode surface. The method of antibody immobilization through streptavidin–biotin interaction is described [16] below.

Materials

Bovine serum albumin (BSA), nitrogen, and distilled water; reduced graphene oxide paper/gold nanoparticle (rGOP/AuNP)-modified electrode; streptavidin (1 mg/mL); wash buffer (10 mM PBS, pH 7.4, with 0.01% Tween 20); and biotinylated anti-*E. coli* O157:H7 antibody (0.5 mg/mL).

Method

1. Drop 10 μL streptavidin onto rGOP/AuNP-modified electrode, and incubate overnight at 4 °C.
2. Rinse the electrode thoroughly with wash buffer.
3. Add 10 μL biotinylated anti-*E. coli* O157:H7, and incubate for 2 h.
4. Wash electrode with distilled water, and dry using a stream of nitrogen.
5. Add 10 μL of 1% BSA, and incubate for 30 min.
6. Wash the electrode again with distilled water, and dry using a stream of nitrogen.

The limit of detection of the immunosensor is 1.5×10^2 CFU/mL when incubated with 1.5×10 to 1.5×10^7 CFU/mL of *E. coli* O157:H7. The developed immunosensor was tested in ground beef and cucumber that has been added with *E. coli* O157:H7 solution with a final concentration ranging from 1.5×10 to 1.5×10^7 CFU/mL. The detection limit of 1.5×10^4 and 1.5×10^3 CFU/mL was reported for ground beef and cucumber, respectively.

Zhang et al. developed a microfluidic exosome-sensing microchip to detect exosomes purified from a colon cancer cell line. The microchip was also able to detect exosomes directly in plasma samples from ovarian cancer patients. The method for antibody immobilization using protein G is described [36] below.

Materials

GO solution (0.5 mg/mL), distilled water, and heating plate; dopamine dissolved in 10 mM Tris buffer (2 mg/mL, pH 9.5); protein G dissolved in PBS (0.2 mg/mL); anti-CD81 antibody (20 μg/mL) and human IgG; and bovine serum albumin (5%).

Method

1. Allow GO to flow onto (3-aminopropyl) triethoxysilane-coated microchip surface using a syringe pump (flow rate: 0.5 μL/min).
2. Allow dopamine to flow onto GO-coated microchip surface using a syringe pump (flow rate: 0.5 μL/min). Perform this step on a 50 °C heating plate for 3 h.
3. Wash with distilled water.
4. Fill the microchip surface with protein G, and allow the protein to react for 16 h at room temperature.
5. Wash with distilled water to remove excess protein.
6. Apply anti-CD81 antibody to protein G-treated surface.
7. Before testing, block the surface with 5% BSA and 1% human IgG for 1 h.

Exosome concentration of $5 \times 10^4/\mu L$ was pumped onto the microchip surface that is coated with and without CD81 antibody. High density of exosomes was attached to the antibody-coated microchip surface, with the exosome size being smaller than 150 nm. Exosome is cell-derived vesicle with the size that ranges from 30 to 150 nm in diameter. The microchip was able to detect up to 50 exosomes/μL, with a 4-log dynamic range. The microchip is also able to quantify exosome concentration from 2 μL of plasma without sample processing.

Our laboratory is broadly interested in the development of immunosensors for pathogen detection. We use screen-printed carbon electrodes with a three-electrode configuration. The working electrode is modified with rGO. Generally, antibodies used for immunosensor are raised against a particular antigen. However, we immobilize normal rabbit IgG using the carbodiimide-assisted amidation reaction onto the electrodes. An immunosensor is currently being developed to detect the non-specific binding of *E. coli* O157:H7. We envision that our work will provide insights into the development of immunosensors for detection of *E. coli*.

4 Conclusion

An antigen–antibody complex serves as the primary principle for immunosensor development. The interaction is detected at the working electrode of an immunosensor in the three-electrode system of an electrochemical cell. The working electrode can be modified through deposition of graphene and its derivatives, and the antibody immobilization can be achieved through several means. Orientation of antibodies is unpredictable, and correct immobilization methods are required to ensure correct antibody orientation for analyte detection and quantification.

References

1. K.S. Novoselov, A.K. Geim, S.V. Morozov, D. Jiang, Y. Zhang, S.V. Dubonos, I.V. Grigorieva, A.A. Firsov, Electric field effect in atomically thin carbon films. Science, **5696**(306), 666–669 (2004). https://doi.org/10.1126/science.1102896
2. A.K. Geim, K.S. Novoselov, The rise of graphene. Nature Mater. **3**(6), 183–191 (2007). https://doi.org/10.1038/nmat1849
3. K.S. Novoselov, A.K. Geim, S.V. Morozov, D. Jiang, M.I. Katsnelson, I.V. Grigorieva, S.V. Dubonos, A.A. Firsov, Two-dimensional gas of massless Dirac fermions in graphene. Nature **438**(7065), 197–200 (2005). https://doi.org/10.1038/nature04233
4. K.I. Bolotin, K.J. Sikes, Z. Jiang, M. Klima, G. Fudenberg, J. Hone, P. Kim, H.L. Stormer, Ultrahigh electron mobility in suspended graphene. Solid State Commun. **146**(9–10), 351–355 (2008). https://doi.org/10.1016/j.ssc.2008.02.024
5. X. Wu, M. Sprinkle, X. Li, F. Ming, C. Berger, W.A. de Heer, Epitaxial-graphene/graphene-oxide junction: an essential step towards epitaxial graphene electronics. Phys. Rev. Lett. **101**(2), 026801 (2008). https://doi.org/10.1103/PhysRevLett.101.026801
6. J. Ito, J. Nakamura, A. Natori, Semiconducting nature of the oxygen-adsorbed graphene sheet. J. Appl. Phys. **103**(11), 113712 (2008). https://doi.org/10.1063/1.2939270
7. J.-C. Charlier, X. Gonze, J.-P. Michenaud, Graphite interplanar bonding: electronic delocalization and van der Waals interaction. EPL (Europhys. Lett.) **28**(6), 403 (1994). https://doi.org/10.1209/0295-5075/28/6/005
8. Y. Zhang, J.P. Small, W.V. Pontius, P. Kim, Fabrication and electric-field-dependent transport measurements of mesoscopic graphite devices. Appl. Phys. Lett. **86**(7), 073104 (2005). https://doi.org/10.1063/1.1862334
9. V. Huc, N. Bendiab, N. Rosman, T. Ebbesen, C. Delacour, V. Bouchiat, Large and flat graphene flakes produced by epoxy bonding and reverse exfoliation of highly oriented pyrolytic graphite. Nanotechnology **19**(45), 455601 (2008). https://doi.org/10.1088/0957-4484/19/45/455601
10. K.L. Choy, Chemical vapour deposition of coatings. Prog. Mater Sci. **48**(2), 57–170 (2003). https://doi.org/10.1016/S0079-6425(01)00009-3
11. G. Faggio, A. Capasso, G. Messina, S. Santangelo, T. Dikonimos, S. Gagliardi, R. Giorgi, V. Morandi, L. Ortolani, N. Lisi, High-temperature growth of graphene films on copper foils by ethanol chemical vapor deposition. J. Phys. Chem. C **117**(41), 21569–21576 (2013). https://doi.org/10.1021/jp407013y
12. W.S. Hummers Jr., R.E. Offeman, Preparation of graphitic oxide. J. Am. Chem. Soc. **80**(6), 1339–1339 (1958). https://doi.org/10.1021/ja01539a017
13. J.W.S. Hummers, Preparation of graphitic acid, in Google Patents (US2798878A) (1957)
14. J. Chen, B. Yao, C. Li, G. Shi, An improved Hummers method for eco-friendly synthesis of graphene oxide. Carbon **64**, 225–229 (2013). https://doi.org/10.1016/j.carbon.2013.07.055
15. M. Choudhary, V. Kumar, A. Singh, M. Singh, S. Kaur, G. Reddy, R. Pasricha, S.P. Singh, K. Arora, Graphene oxide based label free ultrasensitive immunosensor for lung cancer biomarker, hTERT. J. Biosens. Bioelectron. **4**(4), 1–9 (2013). https://doi.org/10.4172/2155-6210.1000143
16. Y. Wang, J. Ping, Z. Ye, J. Wu, Y. Ying, Impedimetric immunosensor based on gold nanoparticles modified graphene paper for label-free detection of *Escherichia coli* O157: H7. Biosens. Bioelectron. **49**, 492–498 (2013). https://doi.org/10.1016/j.bios.2013.05.061
17. X. Mo, Z. Wu, J. Huang, G. Zhao, W. Dou, A sensitive and regenerative electrochemical immunosensor for quantitative detection of *Escherichia coli* O157: H7 based on stable polyaniline coated screen-printed carbon electrode and rGO-NR-Au@ Pt. Anal. Methods **11**(11), 1475–1482 (2019). https://doi.org/10.1039/c8ay02594k

18. J. Chang, S. Mao, Y. Zhang, S. Cui, G. Zhou, X. Wu, C.-H. Yang, J. Chen, Ultrasonic-assisted self-assembly of monolayer graphene oxide for rapid detection of *Escherichia coli* bacteria. Nanoscale **5**(9), 3620–3626 (2013). https://doi.org/10.1039/c3nr00141e
19. L.R. Farris, M.J. McDonald, AFM imaging of ALYGNSA polymer–protein surfaces: evidence of antibody orientation. Anal. Bioanal. Chem. **401**(9), 2821 (2011). https://doi.org/10.1007/s00216-011-5365-9
20. V. Lebec, S. Boujday, C. Poleunis, C.-M. Pradier, A. Delcorte, Time-of-flight secondary ion mass spectrometry investigation of the orientation of adsorbed antibodies on SAMs correlated to biorecognition tests. J. Phys. Chem. C **118**(4), 2085–2092 (2014). https://doi.org/10.1021/jp410845g
21. H. Xu, X. Zhao, C. Grant, J.R. Lu, D.E. Williams, J. Penfold, Orientation of a monoclonal antibody adsorbed at the solid/solution interface: a combined study using atomic force microscopy and neutron reflectivity. Langmuir **22**(14), 6313–6320 (2006). https://doi.org/10.1021/la0532454
22. A.K. Trilling, J. Beekwilder, H. Zuilhof, Antibody orientation on biosensor surfaces: a minireview. Analyst **138**(6), 1619–1627 (2013). https://doi.org/10.1039/c2an36787d
23. L. Cao, *Carrier-Bound Immobilized Enzymes: Principles, Application and Design* (Wiley (Federal Republic of Germany), 2006). https://doi.org/10.1002/3527607668.ch2
24. M. Hartmann, X. Kostrov, Immobilization of enzymes on porous silicas–benefits and challenges. Chem. Soc. Rev. **42**(15), 6277–6289 (2013). https://doi.org/10.1039/c3cs60021a
25. G.T. Hermanson, *Bioconjugate Techniques*, 3rd edn. (Academic Press, London, United Kingdom, 2013). ISBN 9780123822390. https://doi.org/10.1016/b978-0-12-382239-0.00004-2
26. J.H. Jung, D.S. Cheon, F. Liu, K.B. Lee, T.S. Seo, A graphene oxide based immuno-biosensor for pathogen detection. Angew. Chem. Int. Ed. **49**(33), 5708–5711 (2010). https://doi.org/10.1002/anie.201001428
27. V.P. Torchilin, A.V. Maksimenko, V.N. Smirknov, I.V. Berezin, A.M. Klibanov, K. Martinek, The principles of enzyme stabilization. III. The effect of the length of intra-molecular cross-linkages on thermostability of enzymes. Biochimica et Biophysica Acta (BBA)-Enzymol. **522**(2), 277–283 (1978). https://doi.org/10.1016/0005-2744(78)90061-x
28. R. Manjunatha, G.S. Suresh, Melo, J.S. Melo, S.F. D'Souza, T.V. Venkatesha, An amperometric bienzymatic cholesterol biosensor based on functionalized graphene modified electrode and its electrocatalytic activity towards total cholesterol determination. Talanta. **99**, 302–309 (2012). https://doi.org/10.1016/j.talanta.2012.05.056
29. D. Lin, J. Wu, M. Wang, F. Yan, H. Ju, Triple signal amplification of graphene film, polybead carried gold nanoparticles as tracing tag and silver deposition for ultrasensitive electrochemical immunosensing. Anal. Chem. **84**(8), 3662–3668 (2012). https://doi.org/10.1021/ac3001435
30. O. Livnah, E.A. Bayer, M. Wilchek, J.L. Sussman, Three-dimensional structures of avidin and the avidin-biotin complex. Proc. Natl. Acad. Sci. **90**(11), 5076–5080 (1993). https://doi.org/10.1073/pnas.90.11.5076
31. F. Tausig, F.J. Wolf, Streptavidin—A substance with avidin-like properties produced by microorganisms. Biochem. Biophys. Res. Commun. **14**(3), 205–209 (1964). https://doi.org/10.1016/0006-291X(64)90436-X
32. S.-Y. Mao, Biotinylation of antibodies, in *Immunocytochemical Methods and Protocols. Methods in Molecular Biology (Methods and Protocols)* (Springer, New York, USA, 2010). ISBN: 978-1-59745-324-0. https://doi.org/10.1007/978-1-59745-324-0_7
33. M. Shen, J. Rusling, C.K. Dixit, Site-selective orientated immobilization of antibodies and conjugates for immunodiagnostics development. Methods **116**, 95–111 (2017). https://doi.org/10.1016/j.ymeth.2016.11.010
34. C. Wright, K.J. Willan, J. Sjödahl, D.R. Burton, R.A. Dwek, The interaction of protein A and Fc fragment of rabbit immunoglobulin G as probed by complement-fixation and nuclear-magnetic-resonance studies. Biochem. J. **167**(3), 661–668 (1977). https://doi.org/10.1042/bj1670661

35. B. Akerström, L. Björck, A physicochemical study of protein G, a molecule with unique immunoglobulin G-binding properties. J. Biol. Chem. **261**(22), 10240–10247 (1986)
36. P. Zhang, M. He, Y. Zeng, Ultrasensitive microfluidic analysis of circulating exosomes using a nanostructured graphene oxide/polydopamine coating. Lab. Chip. **16**(16), 3033–3042 (2016). https://doi.org/10.1039/c6lc00279j

Nanomaterial for Biosensors

Syed Tawab Shah, Mohammad Khalid, Rashmi Walvekar, and Nabisab Mujawar Mubarak

Abstract Biosensors are briefly discussed in this chapter. Components, properties and types of biosensors are discussed. Nanomaterials having 1D, 2D and 3D materials and their application as biosensors are described. The surface chemistry of biomolecules with nanostructured materials is well explored. Nanomaterials improve selectivity, sensitivity, response rate and reduce the cost of fabrication. Ultrasensitive biosensors with higher efficiency have been fabricated by using carbonaceous materials (CNTs, GO, rGO, CNF), metals, metal oxides, conducting polymers, functionalized materials and composites. The exceptional properties of the nanostructured materials such as surface area, biocompatibility, optical and electrochemical properties have been explored to improve the selectivity and sensitivity of the biosensors.

Keywords Biosensors · Biomolecules · Nanomaterials · Graphene · CNTs · Metal oxide · Conducting polymers · Composite

1 Introduction

A biosensor is an analytical device, used for the detection of a chemical substance, that combines a biological component with a physicochemical detector. Biosensors are a combination of sensitive biomolecules such as DNA, RNA, enzymes, cells,

---

S. T. Shah · M. Khalid (✉)
Graphene and Advanced 2D Materials Research Group (GAMRG), School of Science and Technology, Sunway University, No. 5, Jalan Universiti, Bandar Sunway, 47500 Petaling Jaya, Selangor, Malaysia
e-mail: khalids@sunway.edu.my

R. Walvekar
Department of Chemical Engineering, School of Energy and Chemical Engineering, Xiamen University Malaysia, Jalan Sunsuria, Bandar Sunsuria, 43900 Sepang, Selangor, Malaysia

N. M. Mubarak
Department of Chemical Engineering, Faculty of Engineering and Science, Curtin University Malaysia, CDT 250, 98009 Miri, Sarawak, Malaysia

A. T. Jameel and A. Z. Yaser (eds.), *Advances in Nanotechnology and Its Applications*,
https://doi.org/10.1007/978-981-15-4742-3_3

metabolites oligonucleotides and transducers like optical, electrochemical, piezo-electric, acoustic and calorimetric [1–4]. Biochemical interactions are detected and converted into an electric signal in biosensors which makes it interesting device. Moreover, biosensors are more efficient and can detect minute changes during biological processes occurring due to the interaction of biomolecules. Hence, biosensor devices are fabricated for the diagnosis of various diseases to check food quality and numerous environmental applications. In medical field, biosensors have applications in the detection of pathogens, tumours, biomarkers and toxins effectively at the early stages of disease. Recently, biosensors have grabbed attention due to its fast response time, quantification of the sample at minuscule scale, low manufacturing cost, sensitivity and specificity [5]. The digital signals produced by a biosensor are directly related to concentration of target biomolecules. According to the International Union of Pure and Applied Chemistry definition, "a biosensor is a self-contained integrated device that is capable of providing specific quantitative or semi-quantitative analytical information using a biological recognition element (receptor) which is in direct spatial contact with a transducer element" [6, 7]. The first electrochemical sensor was developed by immobilization of glucose oxidase (GOD) on the oxygen electrode surface for the detection of glucose [8]. Historical development of biosensors is summarized in Table 1. GOD catalyzes the conversion of glucose into gluconic acid which produces two protons and electrons while GOD itself is reduced during this reaction. Reduced GOD reacts with electrons, protons and oxygen to produce hydrogen peroxide and GOD (oxidized). Quantification of glucose can be measured by an increase in H_2O_2 concentration or a decrease in oxygen contents.

A typical biosensor has the following main components:

- A bioreceptor molecule (cells, enzymes, antibodies, etc.) which can detect analyte. In conventional biosensor, bioreceptor molecules (antibodies, cells, enzymes, etc.) can recognize analytes.
- A transducer converts biochemical signals into electronic signals by the interaction of bioreceptor with analytes. The signals produced are either directly or inversely proportional to the concentration of analyte in a sample. Electric potential, phase and intensity of electromagnetic radiations, conductivity, temperature viscosity impedance, etc., could be transducer elements.

A descriptive biosensor is shown in Fig. 1. The transducer generates an output signal which can be digitalized and can be seen on monitor display. Biosensor can be categorized into three generations.

- First-generation biosensors: These biosensors are oxygen-dependent. The electrical signals are produced by the product formed during the reaction.
- Second-generation biosensors: In this type, oxygen is not needed but mediators between transducers and the reaction are used to improve response signal.
- Third-generation biosensors: These biosensors do not require any direct mediators or oxygen. The response signals are mainly from reaction itself. There is direct electron transfer among electrode and biomolecules which excludes other intermediate electron transfer reactions with redox species.

Table 1 Historical development of biosensors

Year	Development	References
1956	Leland C. Clark Jr. Reported the first oxygen electrode. Clark and Ann Lyons developed first glucose enzyme electrode	[10]
1959	The radioimmunoassay (RIA) was established for the detection of hormones with high sensitivity	[11]
1963	First paper on biosensors was published with direct potentiometric detection of urea from urease hydrolysis. The devices were said to be biocatalyst membrane electrode or enzyme electrode at that time	[12]
1967	The enzyme was immobilized in a gel to get first practical enzyme electrode	[13]
1972	Direct electron transfer of cytochrome C was detected at Hg electrode	[14]
1973	Lactate electrode was developed	[15]
1975	First commercial biosensor for diabetic patients was introduced	[16]
1976	First microbial biosensor	[17]
1980	Self-assembled monolayers were used for biosensor applications	[18]
1981	First study of oxidation of NADH at graphitic electrode	[19]
1983	First surface plasmon resonance immunosensor	[20]
1984	First ferrocene-mediated amperometric glucose biosensor	[21]
1988	Electron conducting redox hydrogels were used to elaborate electrical connection of electrodes to redox centres of enzymes	[22]
1988	Immobilized enzymes were used for fast electron transfer	[23]
1990	Mediator modified enzymes were introduced	[24]
1997	Definition of biosensor was introduced by IUPAC in analogy to chemosensor	[6]
2002	Electrodeposition of paint to be used as immobilization matrices. The implanted glucose biosensor worked for 5 days	[25, 26]
2003	Optical biosensors were studied in detail	[27]
2004	Functionalized CNTs were proposed for sensing application	[28]
2007	CNTs FET biosensor	[29]
2009	Microfluidic devices	[30]
2011	Various graphene-based nanocomposites were studied	[31]
	Many metallic nanostructures for SERS in the literature	[32]
2012	Metal oxide-based biosensors	[33]
	Washable wearable biosensor	[34]
2013	Wireless powered contact lens with biosensor	[35]
2014	Implantable biosensor	[36]
	Handheld on-chip biosensing technology	[37]
2015	Biosensor based on smart phone platforms	[38]
2016	Conducting polymers attracted researchers	[39]

(continued)

Table 1 (continued)

Year	Development	References
2017	Biopolymer composites	[40]
2018	Automatic smartphone-based microfluidic biosensor	[41]
2019	Optical fibre biosensor based on SPR	[42]
2020	The nanozyme-enzyme lactate biosensor with twice higher sensitivity	[43]

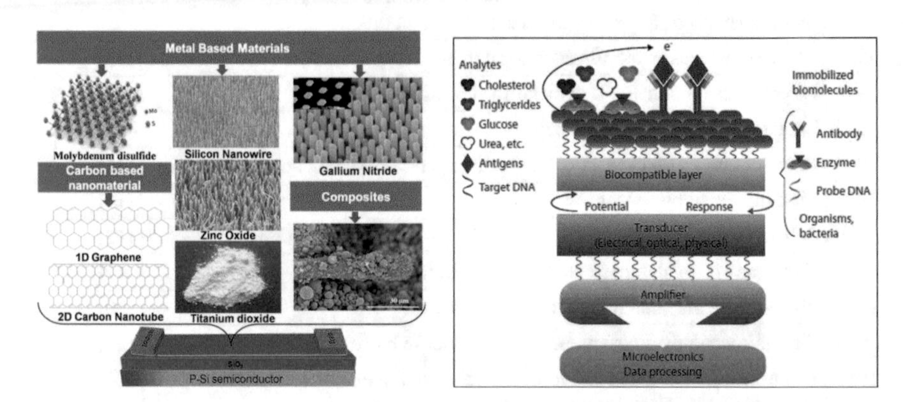

Fig. 1 Nanomaterial-based field-effect transistor (FET) biosensor. Reproduced from Nehra and Singh [9]

- Biosensors can be classified on the type of biorecognition element and the type of transducers used to fabricate biosensors. Major components of a biosensor are shown in Fig. 2. Based on transducers, a biosensor can be electrochemical, optical, calorimetric, piezoelectric and thermoelectric. Based on biorecognition elements, a biosensor is classified as DNA or RNA biosensor, enzymatic, microbial biosensors, whole cell sensor, immunosensors, etc.

2 Carbonaceous Material for Biosensors

2.1 Carbon Nanotubes (CNTs)

Carbon nanotubes (CNTs) are being used in the fabrication of biosensors due to its compatibility with biomolecules both chemically and dimensionally. These features facilitate the electron transfer between the electrode surface and biomolecules [44] as well as catalyzing the reaction [45].

CNTs are extensively used for biosensor using electrochemical and optical techniques. In CNTs-based electrochemical biosensor, CNTs modified working electrode with enhanced electroactive surface area detects the desired analytes. A variety of CNTs-based electrochemical biosensors were fabricated by amending

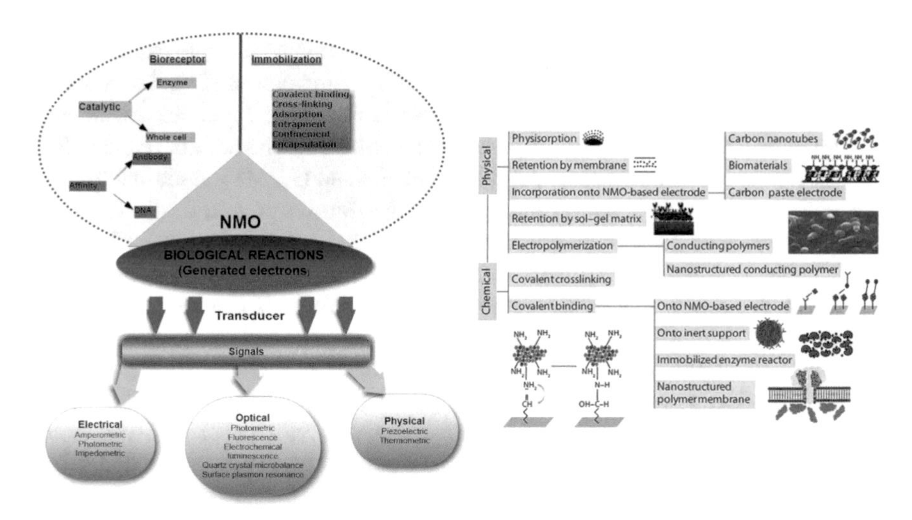

Fig. 2 Main components and types of biosensors

peptides, aptamers, DNA, antibodies, etc., for protein detection and their interaction of biomolecules [46, 47]. CNTs-based FETs have been used for detecting small molecules such as artificial oligonucleotides (aptamers) comparable with the Debye length.

Polymers coating can be used for the growth of CNTs on the surface of existing electrodes. CNTs-based biosensors provide fast electron transfer reactions. Sensitivity of these sensors is higher due to high aspect ratio and higher surface area of CNTs on the electrode surface. Figure 3 is a schematic representation of three-electrode system for electrochemical studies where CNT electrode acts as working electrode while Pt wire and AG/AgCl are counter and reference electrodes, respectively [48].

CNTs could be a good choice for next-generation biomedical sensors for the detection of prostate-specific antigens at the right time [49]. Prostate-specific antigens (PSA) have been detected by using CNT-based label-free immunosensor. An increase

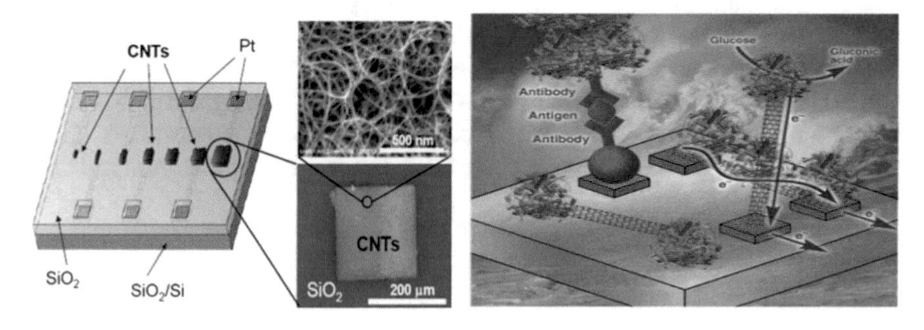

Fig. 3 CNTs-based electrochemical electrodes. Reproduced from Maehashi and Matsumoto [48]

in PSA level in serum has been reported as an indication of prostate cancer with a cut-off limit of 4 ng/mL among prostate hyperplasia and cancer [50]. The amperometric technique has been used for testing cardiac troponin T via electrochemical immunosensor [51]. A biochip with pneumatic micropumps has been developed [52]. This device is a combination of polydimethylsiloxane (PDMS)-based microchannels with pneumatic micropump and amperometric biosensor used for the detection of glucose. A FET using a three-terminal switching device was fabricated using semiconducting single-walled carbon nanotubes (SWCNTs) that were connected to two metal electrodes. The studies revealed that the SWCNTs can be changed from conducting to insulating state by applying voltage to gate electrode [53].

MWCNTs printed electrodes (working, reference and counter) were used for fabricating inkjet printed electrochemical sensor [54]. The device was fabricated by printing MWCNTs ink as a conductive network on a paper, a hydrophobic barrier was used to avoid the absorption of liquid to realize a simple and effective electrochemical sensor. MWCNTs-printed paper sheet resistance was found as 1 kΩ/sq after 33 prints. The fabricated device was cost-effective, easy and disposable paper-based biosensor for the detection of ferrous (Fe^{+2}) ions and dopamine, respectively. Pathogenic viral DNA from hepatitis C virus and genomic DNA of *Mycobacterium tuberculosis* were detected using CNTs-based multiplexed microfluidic sensors [55]. CNTs-based biosensor with fluidic protocol enhanced the limit of detection to femtomolar concentration.

The intrinsic benefits combined with the semiconductor properties, CNTs are widely used in electrochemical biosensors and other biomedical applications. CNTs have been studied for biosensor by conjugation of specific dyes with tagged biomolecules. The reversible quenching of fluorescence of SWCNTs was recently studied using small redox-active organic dye molecules [56]. This device was fabricated by functionalizing CNTs with redox-active dyes, and a suitable reducing agent was used to reverse the quenching. For instance, fluorescence quenching technique was used to detect avidin by using single SWCNTs integrated with dyes and bound to biotin [57]. The tunable near-infrared emission of CNTs can be used to respond to changes in the dielectric function with stable photobleaching. A near-infrared sensor developed for β-D-glucose sensing using CNTs that tuned emission with the adsorption of specific biomolecules by two distinct mechanisms such as fluorescence quenching and charge transfer [58]. The optical biosensor has been fabricated for the detection of DNA hybridization in solution phase by using SWCNT [59]. Combination of marvellous optical, electronic and catalytic features of CNTs with biomolecules opens new vistas for CNTs-based nanocircuitry and bioelectronic devices for biomedical applications. Additionally, CNTs are biocompatible with the human immune system which paves the way to discover cancer therapies and drug delivery systems.

2.2 *Graphene*

Graphene, graphene oxide (GO), reduced GO and functionalized GO have been used as biosensors [60, 61]. Pristine graphene has higher conductance but its application as a biosensor is limited due to its non-functionality towards biomolecules. It is important to understand the chemistry of graphene, and its interaction with biomolecules plays a vital role in applying graphene as a nano-scaffold biosensing [62]. GO sheets have hydroxyl (–OH) and carboxyl (–COOH) groups and a specific two-dimensional structure; due to outstanding properties of GO such as chemical reactivity, optical, thermal, mechanical and electronic properties, it is suitable for medical devices and biosensor application [63]. The conductance of GO depends upon its atomic and chemical structures (Fig. 4) because of structural disorders due to more sp^3 carbons. GO-based films are insulating with a sheet resistance of 10 Ω/sq which is due to sp^3 C–O bonds which results in a decrease in electron transfer and disrupting percolation pathways with sp^2 clusters. Reduction of GO to rGo by thermal or chemical treatment can facilitate the transport of carriers which results in Rs by several orders of

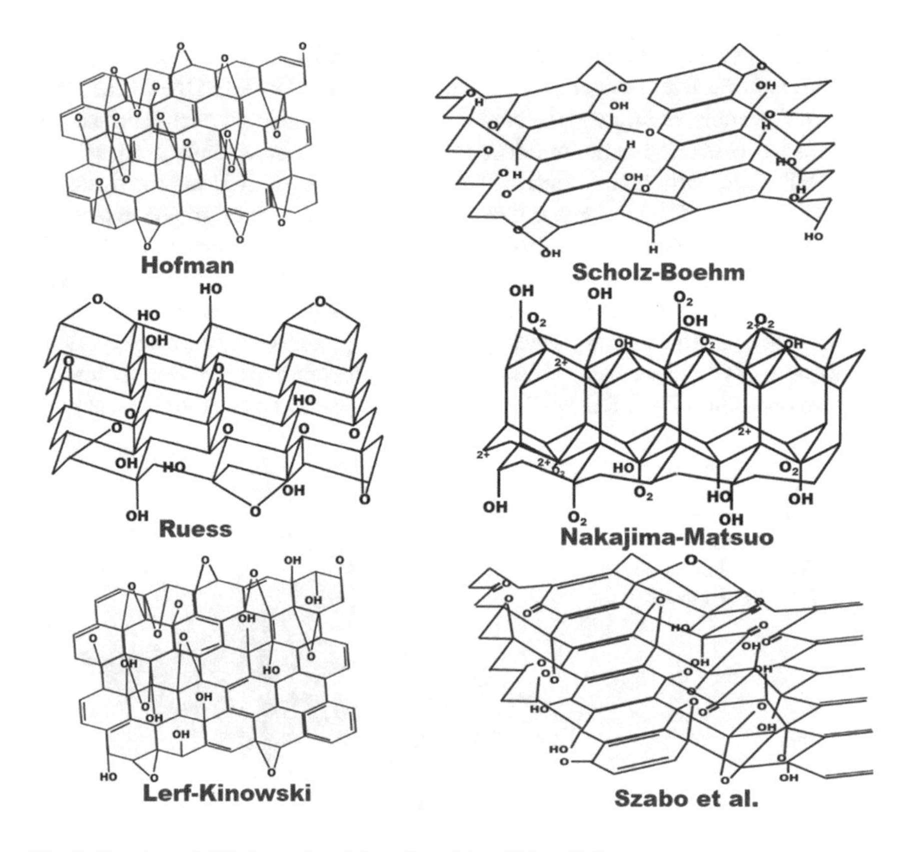

Fig. 4 Structure of GO. Reproduced from Inagaki and Kang [64]

magnitude and hence an insulating GO transforms to semiconductor graphene-like semimetal. The conductance of rGO is found to be 1000 S/m.

The unfunctionalized pristine graphene cannot be dissolved in any solvent, but it can be deposited on substrate by "scotch tape" method which results in uncontrolled size and shape. Chemical vapour deposition (CVD) has been used to get controlled deposition of graphene on a substrate. He et al. [65] used polyethylene terephthalate substrate to fabricate graphene-based FET as shown in Fig. 5. This GFET device operated in an aqueous environment for the detection of bio-species. Both source and drain were insulated to eliminate current leakage due to ionic conduction, and silver/silver chloride or platinum is immersed in solution. The electric double-layer capacitance (EDLC) rises at the interface of channel and electrolyte when the gate potential is applied. In GFET device, there are two chief sensing mechanisms of electrostatic gating effect and the doping effect which have been discovered. In brief, at the gate, when the charged molecules are absorbed on the surface of graphene, they act as additional gating capacitance to change the graphene channel conductance. While in case of doping effect, a direct charge transfer may take place among biomolecules and the graphene channel. In GFER devices, functionalization of graphene can be done via simple electrostatic interaction or p-p interaction rather than covalent bonding. For the development of biosensor, the graphene can be functionalized with the aid of proteins, single-strand DNA (ssDNA), enzymes, etc. As an example, to improve the biostability and specificity graphene surface was modified by incubation in the presence of phosphate-buffered saline using ssDNA [66]. FET sensor with higher sensitivity has been developed [67] to investigate the electrogenic cells. The sensor was developed by the interface of graphene cellular. The functionalized graphene contains the oxygen functional groups in the form of carboxyl, hydroxyl, or epoxy which provide potential benefits to GO sheets in its applicability to use in numerous technological fields. There is a popular method named as coupling mechanism which is used to link proteins with the carboxyl functional group of GO sheets using N-Ethyl-N-(3-dimethylaminopropyl)-carbodiimide-N-hydroxysuccinimide (EDC-NHS). This method is used to make the covalent bond

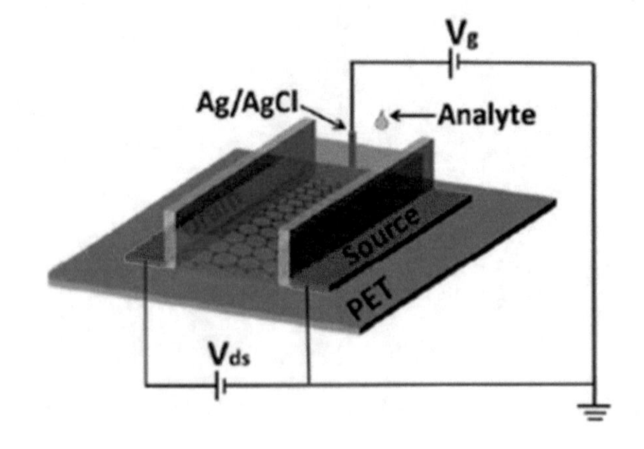

Fig. 5 A solution-gated graphene field-effect transistor on flexible polyethylene terephthalate. Reproduced from He et al. [65]

between the GO sheets and biomolecules which results in enhanced stability and selectivity of the sensing device; this method of making covalent bond is called amid (C–N) by amidation reaction. The antibody conjugate presents on the surface of aminated GO sheets use EDC-NHS chemistry in order to establish an appropriate biointerface which in turn helps in the investigation of lipid–lipid interactions [68]. Although GO has poor electrical conductivity due to –OH and –COOH groups, it still provides excellent electron transfer behaviour as it exhibits well-defined redox peaks via cyclic voltammetry (CV) studies in ferro/ferricyanide ($[Fe(CN)_6]^{3-/4-}$) and hexaammineruthenium(III/II) ($[Ru(NH_3)_6]^{3+/2+}$) analytes, respectively [63]. The square root of the magnitude of redox peak current shows a linear increase via CV analysis at different scan rates suggest that the process is diffusion-controlled. Thus, it confirms that GO has the potential to serve as an electrochemical biosensor. Subsequently, most of the atoms are exposed on the surface of GO sheets; therefore, a light change in adsorption of protein molecules on the surface of GO sheets turns in the form of significant changes in the electrical properties which help in the formation of highly selective a sensitive biosensor. Papakonstantinou et al. used nanoflakes of graphene for the simultaneous sensing of biological molecules (dopamine, ascorbic acid, and uric acid) [69]. The recent reports show the ability of GO to be used in electrochemical sensing of glucose via an interaction between carboxyl acid groups of GO sheets with amines of glucose oxidase [70]. A PDMS paper glass-based microfluidic sensor was developed for the detection of multiplexed pathogen detection (Fig. 6) [71], and this one step detection was demonstrated by aptamer-functionalized GO. Molecular probing in living cell was carried out by using GO sheet and aptamer carboxyfluorescein complex [72].

rGO has improved electrochemical conductance due to less oxygenated functional groups. rGO has a substantial amount of functionality and desired conductance that makes them suitable for biosensor applications. rGO sheets deposited on ITO electrophoretically showed improved conductivity and were used for the detection of aflatoxin B1 in food [73] and immunosensing platform was fabricated using rGO for detecting ultrasensitive antigens [74]. A label-free, immunosensor with high sensitivity, selectivity and reproducibility has been described using anti-apolipoprotein B functionalized mesoporous few-layer rGO and nickel oxide (rGO-NiO) composite for quantifying low-density lipoproteins (Fig. 7) [75]. rGO has been used for developing enzymatic biosensors without mediators to fabricate third-generation amperometric glucose biosensor [76]. rGO and quantum dots (QD) composite have been used for bioimaging of tumour cells with photothermal therapy [77].

2.3 *Carbon Nanofibers*

Carbon nanofibers (CNFs) have been used for biosensor transducer material because of cost-effectiveness and inertness in aqueous medium [78]. A variety of polymers (phenolic resin, polyimides, polybenzimidazole poly(vinyl alcohol), poly(vinylidene fluoride), polyacrylonitrile, and lignin) can be used for the synthesis of CNFs. A

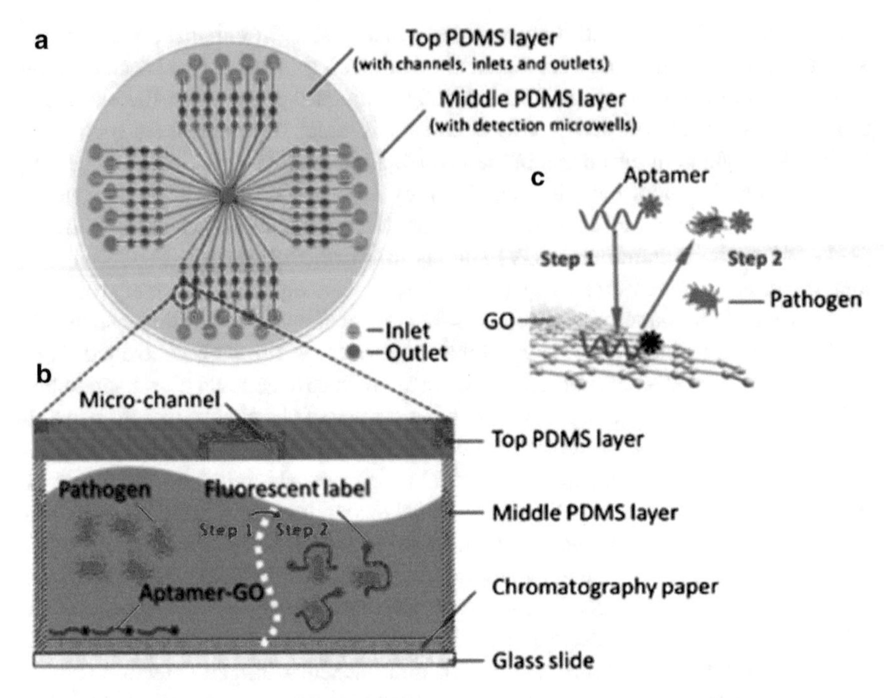

Fig. 6 An aptamer-functionalized GO-based microfluidic biosensor using polydimethylsiloxane (PDMS) paper hybrid for detection. Reproduced from Zuo et al. [71]

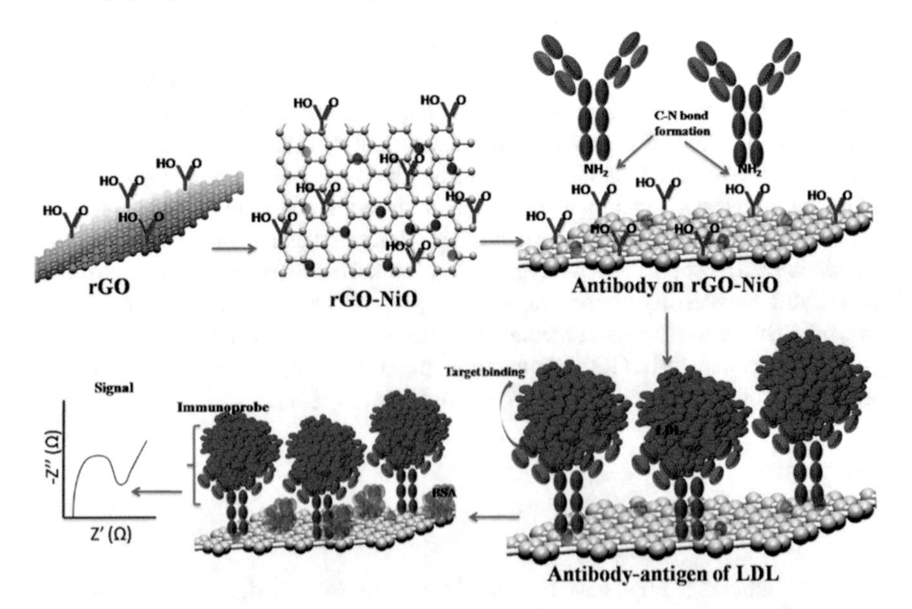

Fig. 7 Functionalized reduced graphene oxide-nickel oxide (rGO)-NiO composite with antibody used for detecting blood low-density lipoprotein (LDL). Reproduced from Ali et al. [75]

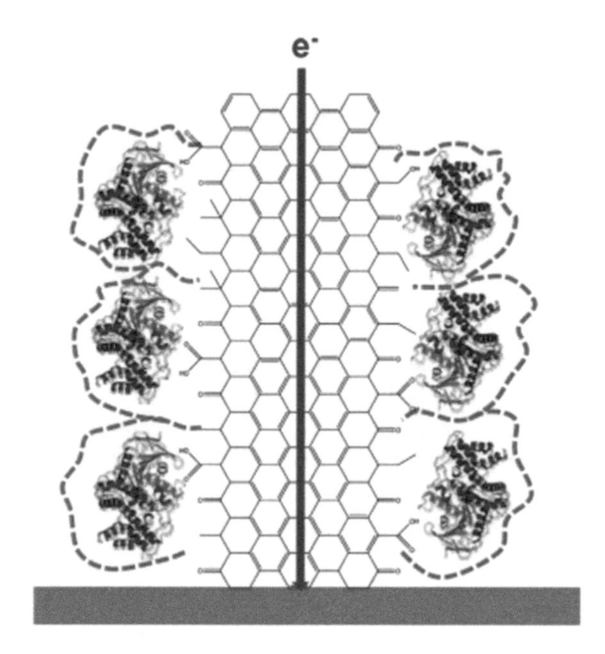

Fig. 8 Electrochemical biosensor using silica nanofibers. Reproduced from Vamvakaki et al. [79]

suitable polymeric material can be electrospun to fibres having a diameter ranging from nanometres to micrometres. Protein molecules can easily be loaded due to high surface area and porosity. A biochemically synthesized silica CNFs nanocomposite-based biosensor was developed where electron transfer was achieved by CNF (Fig. 8) [79]. Various CNFs-based biosensors were used for detecting biomolecules such as glucose [78], proteins [80] and *Escherichia coli* [81].

3 Metal-Based Biosensors

Metal and metal oxides have received much consideration in recent years in the field of biosensor applications because of its marvellous chemical, electrochemical and optical properties [82, 83]. Metal and metal oxides are important for biosensor applications because of immobilization of biomolecules with improved conformation, orientation and outstanding biological activity. Metal and metal oxide nanoparticles are recognized to be dynamic materials due to high surface area for immobilization of biomolecules with improved orientation, conformation, and outstanding bioactivity which results in enhanced performance of a biosensor. The performance of metal oxide-based biosensors can be amended by modifying the bio interface between nanoparticles and biomolecules by changing functionality, morphology and size [84].

Various metal-based biosensors with excellent stability and selectivity have been fabricated with many noble metals such as Au [85, 86], Ag [87], Pt [88], Pd [89] and Ni [90, 91].

Metal/metal oxide-based nanoparticle-based biosensor devices have been fabricated such as biosensors for cell tracking, in vivo sensing, therapy monitoring and point-of-care (POC) diagnostics devices. Various sensing platforms have stimulated using spectral and optical properties of nanoparticles. Diverse sizes and morphologies of nanoparticles can be used to fabricate biosensor for a specific analyte (Fig. 9) [92]. The marvellous properties can be manipulated precisely by changing the composition of nanoparticles [93].

Various techniques have been used to prepare noble metal nanoparticles (mNPs) such as chemical reduction, hydrolysis, thermal, chemical/physical vapour deposition, laser ablation, grinding, etc., majority of these mNPs exhibit size-related properties as compared to bulk material.

Silver and gold nanoparticles are mostly studied materials for biosensors and other biomedical applications [94]. Silver and gold nanoparticles show localized surface plasmon resonance (LSPR) properties that can be used to fabricate new

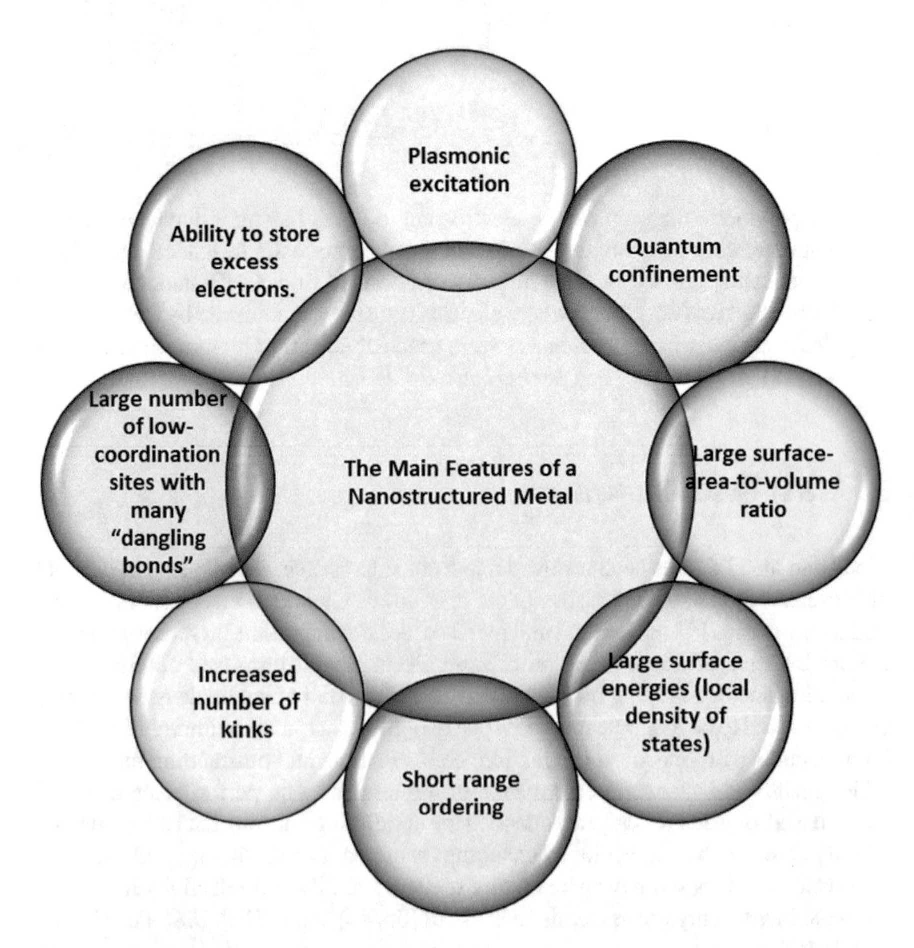

Fig. 9 Main characteristics of nanostructured metals

optical biosensors. The mNPs produce a high and scattering due to the oscillation of conduction electrons. The LSPR of Au and Ag nanoparticles have high absorption coefficients and scattering in UV-visible wavelength which can be detected much better than organic dyes due to high sensitivity [95].

Au nanoparticles have been widely explored due to their tuneable optical properties which makes it suitable for the detection of biomolecules [96]. Moreover, Au NPs can enhance Raman and Rayleigh signals to detect biomolecules. Colloidal Au NPs show vibrant because of surface plasmon resonance absorption. Affinity-based biosensors have been reported in the literature (Fig. 10) [92]. Hybridized DNA has been detected by using fluorescence biosensors [97].

Silver nanoparticles (Ag NPs) are extensively used for developing electrochemical and optical biosensors [98]. Highly sensitive LSPR-based optical sensor was fabricated by using triangular Ag NPs [99]. Ag nanoparticles composite with CNTs can enhance electrochemical properties; hence, mediator-free biosensors can be developed. For instance, H_2O_2 biosensor was established by myoglobin immobilized Ag NPs enabled CNTs film [100]. Chitosan and Nafion conjugated with Pt and Pd nanoparticles have been used for fabrication of biosensor for the detection of glucose and cholesterol in serum [101, 102].

Metal oxide nanoparticles are materials of interest for the development of biosensors [83]. Unique properties of metal oxide nanoparticles can make interface with

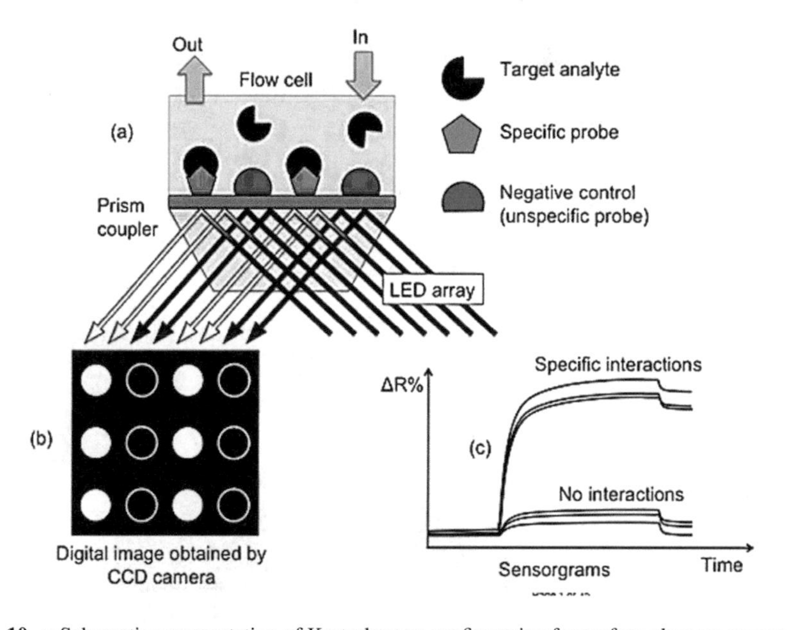

Fig. 10 **a** Schematic representation of Kretschmann configuration for surface plasmon resonance. **b** Data are recorded as intensity variation of the reflected light at a fixed angle. **c** Corresponding sensorgrams for the specific interactions of the analyte with the spots on the surface. Reproduced from Scarano et al. [92]

bioreceptor molecules for biorecognition and can be used to develop novel bioelectronic devices with unique functions [83]. Metal oxides can be classified as organic, inorganic and inorganic–organic composites [103]. Inorganic metal oxides such as ZnO, NiO, CeO_2, SiO_2 and TiO_2 Fe_2O_3 can be potential candidates for biosensor applications because of their multifunctional properties [82].

Figure 11 portrays the salient features of metal oxides that are suitable for the development of biosensor. These metal oxide nanoparticles can be used to fabricate biosensor with high sensitivity, low limit of detection for the quantification of analyte and are cost-effective. It is necessary to understand the carrier transport mechanism for the fabrication of an electrochemical transducer. The knowledge about electron transfer properties of metal oxides gives better information on carrier transport mechanisms for fabricating electrochemical transducer. Metal oxide nanoparticles show inertness and have a high surface to volume ratio, high surface reactivity biocompatibility, stability and high electron transport properties [82, 83]. Nevertheless, functionalization of these metal oxides with organic moieties is a bit challenging to get novel biosensors [104]. A small mediator species provides a shuttle between electrode and

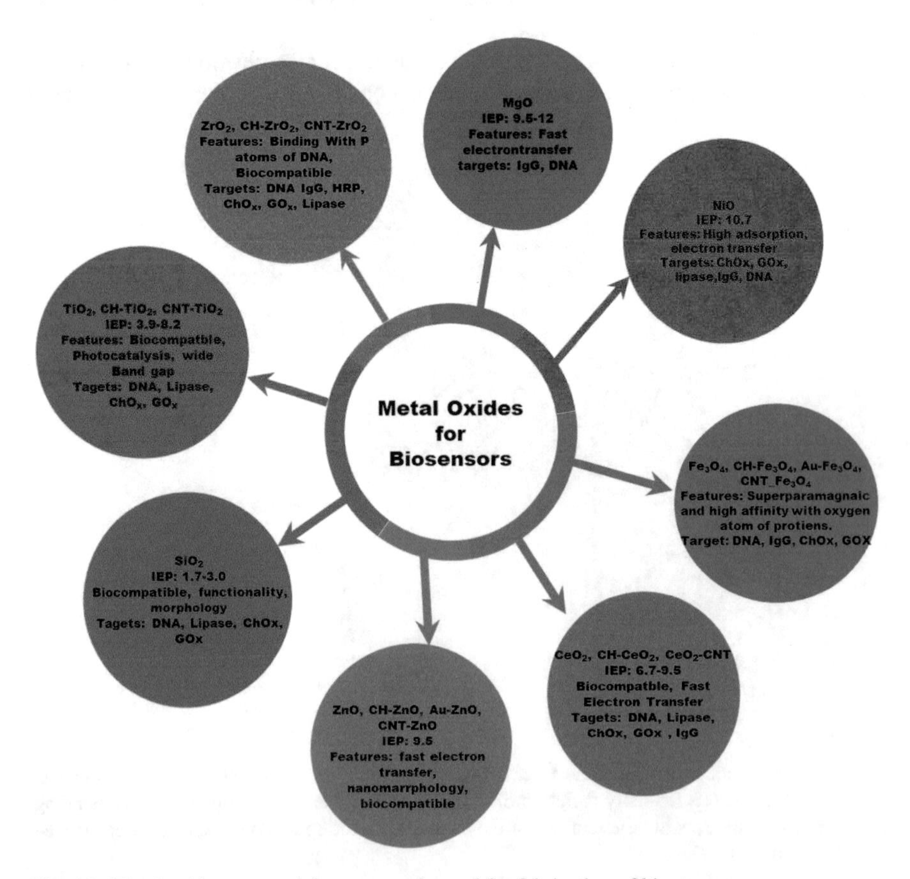

Fig. 11 Metal oxide nanoparticles commonly used for fabrication of biosensor

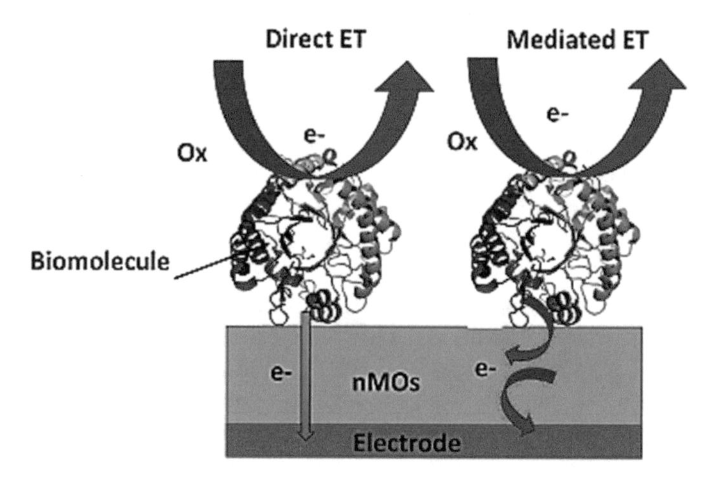

Fig. 12 Schematic diagram of electron transfer mechanism on the surface of electrode. nMOs, nanostructured metal oxides; Ox, oxidation. Reproduced from Malhotra et al. [33]

bioreceptor. Active site of bioreceptor molecules can be activated by metal oxides which can be detected electrochemically by ET. The active sites of bioreceptors are usually hidden in the molecular structure which hinders the direct ET [105]. Direct ET is achieved by modifying the electrode surface with metal oxides which acts as mediators and establishes a functional bio interface (Fig. 12). Biomolecules can be denatured and lose bioactivity if absorbed on bulk materials. However, CNTs and chitosan biopolymer composite with metal oxide nanoparticles can act as biocompatible transducers and biomolecules can retain their bioactivity [33]. The bio-interface between biomolecules and metal nanoparticles play key role as morphology, surface area, etc., can modify the efficiency of a biosensor. An efficient bio interface can retain bioactivity and shows enhanced stability due to biocompatible microenvironment with fast electron transfer.

3.1 Functionalized Metal Nanoparticles

Functionalized metal nanoparticles have been widely used for various applications including biosensors, catalysis and photonics [106–108]. For instance, gold nanoparticles have been studied for the development of biosensors because of their good optical properties and good oxidative stability. Gold nanoparticles are good vehicles for tracers while silver nanoparticles have an excellent light scattering and absorbing [85, 86, 109]. These nanoparticles can be functionalized with carboxylic acids, nitriles, phosphine, amines [110], disulphides [111], thiol [112] groups covalently [113].

Sulphur has a strong affinity to the metallic surface and metal sulphur bond is strong which provides immobilization of thiol group. Organosulphur can easily interact with metal nanoparticles such as Fe, Cu, Au, Ag Pt and Pd by conjugation. The chemical adsorption energy of metal–sulphur bond is 126 kJ mol^{-1}, and hence, sulphur compounds replaces solvent molecules and can be grafted on the metal surface. Sulphydryl (RSH) group in thiols can easily attach to the metal surface. Disulphides and thiols can be chemically adsorbed on the surface of metals. Grafted molecules can be used to modify the properties of metal nanoparticles.

Functionalization of metal nanoparticles with amines stabilizes the metal nanoparticles. For example, hexadecylamine-functionalized Pd nanoparticles show higher dispersion stability of nanoparticles [114]. However, thiol interaction with metal nanoparticles is stronger than amines. Organosulphur functionalized metal nanoparticles are smaller in size than amine functionalized nanoparticles. Tetraalkylammonium halides are generally used for stabilizing noble metal nanoparticles. Au nanoparticles can be functionalized to get self-assembled monolayer on the surface of gold. Thiolated layer can be formed by using various techniques.

Thiol functionalized gold nanoparticles can be directly attached to biomolecules such as RNA and DNA (Fig. 13). Figure 14 depicts a sensor for colorimetric detection of adenosine by using gold functionalized with 30 thiol modified DNA (30 AdapAu), 50 thiol modified DNA (50 AdapAu) and linker DNA [115].

The functionalization of metal nanoparticles with different ligands indicates the multifunctional adsorption. For example, Au nanoparticles functionalized with bifunctional group (N-isobutyryl-L-cysteine) and a trifunctional group, penicillamine, with amine, thiol and carboxylate interaction on the surface can impart stability to nanoparticles. This functionality aids the metal nanoparticles to proteins or antibodies by amide (C–N) linkage.

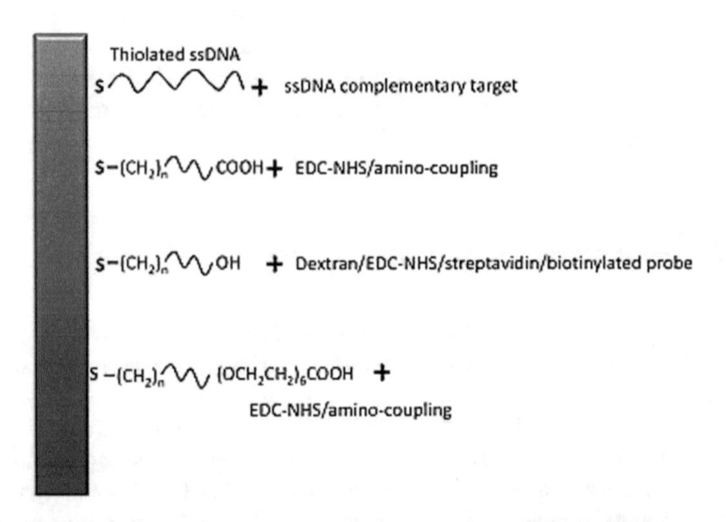

Fig. 13 Various immobilization methods using chemical linkers. A self-assembled monolayer (SAM) is formed as the foundation of the array comprising alkanethiols-containing terminated amine, hydroxylic, or carboxylic functional groups. EDC, 1-Ethyl-3-(3-dimethylaminopropyl)-carbodiimide; NHS, N-hydroxysuccinimide. Reproduced from Scarano et al. [92]

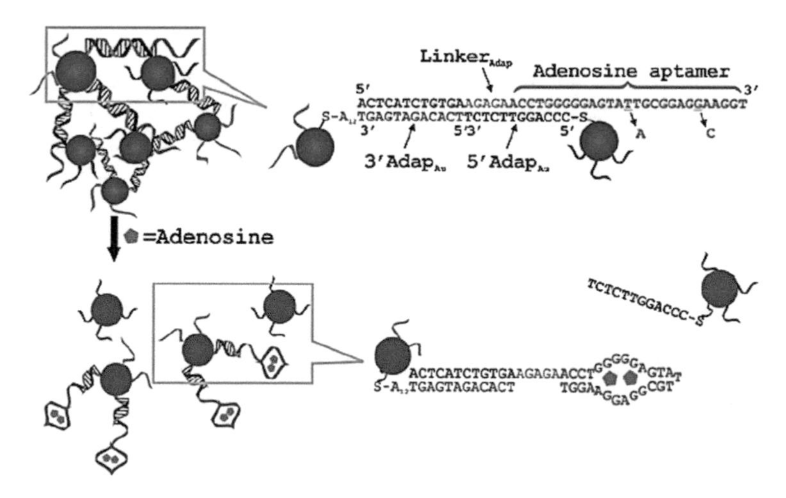

Fig. 14 Graphical illustration of adenosine detection by colorimetry. Reproduced from Liu and Lu [115]

A low-density lipoprotein sensor was reported by attaching L-cystine capped quantum dot to good nanoparticles [116]. Phosphine functionalized metal nanoparticles are also reported in the literature, but their applications are limited due to their stability. The stability can be improved by used polyphosphine ligands [113].

3.2 Functionalized Metal Oxides

Metal oxides have various properties such as catalytic, magnetic, UV absorption and fluorescence. They have numerous applications due to its interesting properties, stability and low cost. Metal oxides are widely used in field of ceramics, coatings catalysis and interacting with biomolecules. Metal oxide-based sensors and biosensors have been studied in the last few decades.

Metal oxides can serve as a signal transduction element and a target recognition element for sensors. Optical, magnetic and electric bases sensing has been used for signal transduction. The properties on these metal oxides can be modified by functionalizing it with various moieties. Carboxylic acids, amines and thiols are used to functionalize metal oxides. However, amine and thiol functionalization are rarely used for metal oxides. For example, tin oxide (SnO_2) is functionalized using thiourea [117]. Silanes and phosphonates are commonly used to modify the surface of metal oxides. Usually, the ligand exchange is carried out for metal oxides without rearrangements by using the same type of (polydentate) ligands to keep identical coordination sites. Metal oxide surfaces can be modified by using carboxylate ligands. For instance, TiO_2 nanoparticles were modified by carboxylic groups and investigated by a surface-binding transient probe molecule wherein the excited triplet state probe molecule was produced by photo-induced interfacial charge recombination [118]. The dispersion stability of Fe_2O_3 was improved by using carboxylic acids (pH 7) by shifting the isoelectric point of the coated NP to pH 3.0 [119]. Additionally, silane moieties are broadly used for the attachment of organic molecules by Si-O-M bonding several alkoxysilanes and silanes have been used to modify metal oxides surface such as Titania, alumina zirconia, vanadium pentoxide, silicon dioxide and tin oxide. Through several functional capabilities of silanes (including epoxies, amino, carboxylic acid, cyano, etc.), which can easily interact with metal oxide surfaces. Silanes can be attached to the metal oxide surface by adding it during metal oxide synthesis. Petoral et al. explained the formation of SAMs of mercaptopropyltrimethoxysilane on ZnO surface [120]. Silicon and silicon dioxide were functionalized by using 3-aminopropyltriethoxysilane as coupling agent by surface hydroxylation. However, surface modification through multistep chemical treatment on the surface of biosensor may result in device complexity. To solve this problem, biomolecules are immobilized on metal oxides surface through electrostatic interaction.

4 Nanocomposite

Nanocomposites are nanomaterials where one or more phases are nano-sized (0D, 1D, 2D) embedded in metal, ceramic or polymeric material [121, 122]. Novel properties can be achieved by using various compositions of organic and inorganic components [121, 122]. Composites are fabricated by using two materials by weak bondings such as hydrogen bonding, weak electrostatic interactions, covalent bonds and van der Waals forces [123]. Nanocomposites are fabricated by using nanomaterials as one of the components [121–123]. Various nanoparticles, CNTs, nanowires, nanorods and nanofibers have been used to fabricate nanocomposites. These nanocomposites show synergy of properties of individual components. These properties can be tuned by adding two or more nanomaterials or by changing their compositions. The multifunctional properties such as redox reactivity catalytic activity, electrical conductivity high mechanical strength and surface-to-volume ratio can be utilized to fabricate biosensors. Enzyme-based nanobioelectronic devices have been fabricated using nanocomposites [124, 125]. Additionally, nanocomposites are widely used as drug delivery systems gas sensors, batteries and artificial implants [126–129]. Nanocomposites have been used for various biomedical applications such as cancerous cells [130] and pathogens [131].

Plethora of nanocomposite including metal-metal oxides, metal-carbon, metal oxide-polymer has been used for fabrication of enzymatic biosensors [40, 132]. Various chemical routes such as coprecipitation, hydrothermal, sol-gel and self-assembly process can be used to synthesize nanocomposites [133]. Nanocomposites can act as transducers for immobilization of biomolecules by physical or chemical interaction or entrapment [134]. Nanocomposites can be inorganic nanomaterials, organic nanomaterials, organic–inorganic hybrid nanomaterials.

Carbon nanotubes, gold, silver, platinum, palladium, titania, alumina, zirconia, ceria, magnetite and zinc oxide nanoparticles are inorganic materials used in nanocomposite by combing two or more elements. Biosensors with higher efficiency can be fabricated by using organic–inorganic nanocomposite. Polyaniline, polypyrrole, poly(3,4-ethylenedioxythiophene), chitosan and other organic polymers can be added to get inorganic–organic hybrid nanocomposites.

Chitosan-graphene electrode was reported to show 1.12×10^{-9} mol/cm^2, with a wider detection range (0.08–12 mM) for glucose having a low limit of detection 0.02 mM, and higher sensitivity (37.93 A/mM cm^2) due to direct electron transfer [135].

CdS-CNT nanocomposite was used for developing enzymatic biosensor by using electrochemiluminescence for the detection of choline and acetylcholine with high sensitivity [136]. Iron oxide–chitosan nanocomposite was immobilized with tyrosinase enzyme used to develop biosensor for the detection of phenolic compounds. Iron oxide has high surface area while chitosan has high porosity which resulted in a higher amount of enzyme loading and enzyme could retain bioactivity [137].

4.1 Hybrid Polymer Nanocomposites

Hybrid nanocomposites of polymers such as metal oxides with chitosan and conducting polymers and carbonaceous material have been used for fabrication of biosensor devices [89, 132, 138]. Macromolecules having delocalized electrons in their backbone show good solubility and electronic properties. Due to easy synthesis procedure polymer-based nanocomposites and excellent charge transfer properties, these materials are suitable for the development of biosensors. In vitro and in vivo sensing of glucose using Au NPs and PANI nanocomposite was reported with higher sensitivity [124]. CNTs and PANI hybrid nanocomposites were synthesized by in situ polymerization using SWCNTs as a template [139]. This nanocomposite showed higher electrical conductance with enhanced carrier mobility and Seebeck coefficient as compared to PANI alone [139]. Fe_3O_4-PPY-Au biocompatible nanocomposite with high magnetism was fabricated to detect ascorbic acid [140]. Bandgap of graphene-polymers can be tuned [87, 140]. These nanocomposites have synergy and are suitable for biosensor applications. Graphene–polyvinylpyrrolidone–PANI nanocomposite was used to fabricate cholesterol biosensor [141].

Hybrid nanocomposites of chitosan with inorganic moieties are nontoxic, biocompatible, cost-effective, highly permeable and easily available [142]. Chitosan has amino group which can be easily modified to get chitosan derivatives. This modification can provide hydrophilic environment for biomolecules. Chitosan-magnetite nanocomposite was used to fabricate biosensor for detecting ochratoxin A via the electrochemical technique [143]. A very good enzyme activity and biocompatibility were reported for ZnO-chitosan biosensor for the detection of cholesterol [144].

4.2 Metal Oxide Nanohybrid Materials

Metal oxides-CNTs nanohybrid materials have been studied for the biosensor application [145]. Metal oxide-CNTs nanohybrid materials provide excellent biosensing properties. CNTs have high oxidation stability, high aspect ratio, large surface area, and chemical inertness makes it a suitable candidate for nanohybrid materials to get nanobiointerface with biomolecules and higher efficiency. Various techniques can be used to synthesize CNTs nanohybrids with inorganic and organic species. These heterogeneous structures show synergism, i.e. combined properties of bot nanomaterials used; due to their superior electrochemical properties, these heterostructures can be the best materials for electrochemical biosensors. A few examples of such nanohybrids are TiO_2-CNTs, NiO-CNTs, Fe_3O_4-CNTs, Co_3O_4-CNTs, ZnO-CNTs, CeO_2-CNTs, ZrO_2-CNTs, etc. Mercepto or amine-terminated metal oxides can anchor to the carboxylic groups on the CNTs via amide bond formation [146].

Metal oxides can be attached directly to the surface of carboxylated CNTs because of hydrophilic nature; for instance, TiO_2, MgO and MnO_2 were attached to carboxylated CNTs [147–149]. To get a uniform distribution of metal oxides on CNTs, surface capping agent can be used. For example, 2-amino-ethylphosphoric acid-modified CNTs can be attached with TiO_2 nanoparticles by using cetyltrimethylammonium bromide as a capping agent [150] where well-distributed phosphine groups on CNTs can attach titania by creating a driving force. Similarly, alkoxysilane-modified CNTs were decorated with SiO_2 nanoparticles [150].

Fe_3O_4-carboxylated CNTs were used [151] to immobilize acetylcholinesterase via covalent interactions for the detection of organophosphorus insecticide. The fabricated biosensor showed low limit of detection (0.1 nM), high sensitivity (0.475 mA/mM), good stability and reusability of more than 2 months [151]. Zirconia, Zirconia-CNTs-based nucleic acid biosensor was fabricated to detect *Mycobacterium tuberculosis* [152]. Selective detection of dopamine was reported by using SnO_2 and ZnO biosensors [153]. Leukaemia cell detection was reported using titania-CNTs composite [154]. Additionally, various metal oxide-CNTs such as platinum and gold have been used for immobilization of biomolecules and selective conjugation for developing nanobioelectronic sensors. Platinum-CNTs nanocomposite was used for the detection of biomolecules [155].

Graphene-based (GO, rGO, N-doped GO) was studied to be used as scaffolds for growing metal oxide nanoparticles for biosensing applications [156]. GO and rGO have been extensively used for the fabrication of biosensors due to its suitable properties [157]. Researchers are putting their efforts to improve graphene-based biosensors. A sheet-like structure of graphene and graphene oxide is feasible for metal or metal oxide nanoparticles as well as immobilization of biomolecules. Edges of graphene sheets are more reactive towards inorganic and organic moieties as compared to basal planes. Metal oxide-graphene nanocomposite plays an important role in biosensor fabrication. The single-layered GO shows excellent redox behaviour of $[Fe(CN)_6]^{-3}$ mediator for the detection of glucose concentration via the cyclic voltammetry technique [158]. A biosensor for the detection of glucose oxidase was fabricated by using ZnO-rGO nanocomposite [159]. The direct electron transfer is due to nanostructured ZnO wiring among electrode and redox centre of enzyme [159].

5 Conclusions

In the past few years, the biosensor application of materials is well explored and is still ongoing for the fabrication and development of new sensing devices with higher sensitivity and selectivity. Nanomaterials are promising candidates for biosensor devices due to their excellent properties such as biocompatibility and higher surface-to-volume ratio. This chapter covers the nanomaterials used to devise biosensors with higher sensitivity, biocompatibility, improved selectivity and cost-effectiveness.

6 Conflict of Interest

The authors of this chapter have no conflict of interest.

References

1. S. Pandit et al., Nanotechnology based biosensors and its application. Pharm. Innov. **5**(6, Part A), 18 (2016)
2. P. Mehrotra, Biosensors and their applications—a review. J. Oral Biol. Craniofac. Res. **6**(2), 153–159 (2016)
3. A. Solaimuthu et al., Nano-biosensors and their relevance in tissue engineering. Curr. Opin. Biomed. Eng. **13**, 84–93 (2020). https://doi.org/10.1016/j.cobme.2019.12.005
4. F. Bettazzi et al., *Biosensors and Related Bioanalytical Tools, in Past, Present and Future Challenges of Biosensors and Bioanalytical Tools in Analytical Chemistry: A Tribute to Professor Marco Mascini* (2017), pp. 1–33
5. A. Turner, I. Karube, G.S. Wilson, *Biosensors: Fundamentals and Applications* (Oxford University Press, 1987)
6. D.R. Thévenot et al., Electrochemical biosensors: recommended definitions and classification. Anal. Lett. **34**(5), 635–659 (2001)
7. M. Chakraborty, M.S.J. Hashmi, An overview of biosensors and devices, in *Reference Module in Materials Science and Materials Engineering* (2017)
8. J. Wang, Glucose biosensors: 40 years of advances and challenges. Electroanalysis **13**(12), 983–988 (2001)
9. A. Nehra, K.P. Singh, Current trends in nanomaterial embedded field effect transistor-based biosensor. Biosens. Bioelectron. **74**, 731–743 (2015). https://doi.org/10.1016/j.bios.2015.07.030
10. L.C. Clark Jr. et al., Continuous recording of blood oxygen tensions by polarography. J. Appl. Physiol. **6**(3), 189–193 (1953)
11. R. Yalow, S. Berson, Immunoassay of endogenous plasma insulin in man. 1960. Obesity Res. **4**(6), 583 (1996)

12. S. Katz, G. Rechnitz, Direct potentiometric determination of urea after urease hydrolysis. Fresenius' J. Anal. Chem. **196**(4), 248–251 (1963)
13. S. Updike, G. Hicks, Reagentless substrate analysis with immobilized enzymes. Science **158**(3798), 270–272 (1967)
14. S.R. Betso, M.H. Klapper, L.B. Anderson, Electrochemical studies of heme proteins. Coulometric, polarographic, and combined spectroelectrochemical methods for reduction of the heme prosthetic group in cytochrome C. J. Am. Chem. Soc. **94**(23), 8197–8204 (1972)
15. P. Racine, W. Mindt, On the role of substrate diffusion in enzyme electrodes, in *Biological Aspects of Electrochemistry* (Springer, 1971), pp. 525–534
16. J. Wang, Electrochemical glucose biosensors. Chem. Rev. **108**(2), 814–825 (2008)
17. M. Hikuma et al., A potentiometric microbial sensor based on immobilized *Escherichia coli* for glutamic acid. Anal. Chim. Acta **116**(1), 61–67 (1980)
18. R.G. Nuzzo, D.L. Allara, Adsorption of bifunctional organic disulfides on gold surfaces. J. Am. Chem. Soc. **105**(13), 4481–4483 (1983)
19. H. Huck, H.L. Schmidt, Chloranil as a catalyst for the electrochemical oxidation of NADH to NAD+. Angew. Chem. Int. Ed. Engl. **20**(4), 402–403 (1981)
20. B. Liedberg, C. Nylander, I. Lunström, Surface plasmon resonance for gas detection and biosensing. Sens. Actuators **4**, 299–304 (1983)
21. A.E. Cass et al., Ferrocene-mediated enzyme electrode for amperometric determination of glucose. Anal. Chem. **56**(4), 667–671 (1984)
22. E. Csoeregi, D.W. Schmidtke, A. Heller, Design and optimization of a selective subcutaneously implantable glucose electrode based on "wired" glucose oxidase. Anal. Chem. **67**(7), 1240–1244 (1995)
23. T. Ikeda et al., Direct bioelectrocatalysis at electrodes modified with D-gluconate dehydrogenase. Agric. Biol. Chem. **52**(10), 2655–2658 (1988)
24. P. Bartlett, V. Bradford, R. Whitaker, Enzyme electrode studies of glucose oxidase modified with a redox mediator. Talanta **38**(1), 57–63 (1991)
25. C. Kurzawa, A. Hengstenberg, W. Schuhmann, Immobilization method for the preparation of biosensors based on pH shift-induced deposition of biomolecule-containing polymer films. Anal. Chem. **74**(2), 355–361 (2002)
26. R.L. Weinstein et al., Accuracy of the 5-day FreeStyle Navigator Continuous Glucose Monitoring System: comparison with frequent laboratory reference measurements. Diabetes Care **30**(5), 1125–1130 (2007)
27. R. Ulber, J.-G. Frerichs, S. Beutel, Optical sensor systems for bioprocess monitoring. Anal. Bioanal. Chem. **376**(3), 342–348 (2003). https://doi.org/10.1007/s00216-003-1930-1
28. B.J. Hinds et al., Aligned multiwalled carbon nanotube membranes. Science **303**(5654), 62–65 (2004). https://doi.org/10.1126/science.1092048
29. B.L. Allen, P.D. Kichambare, A. Star, Carbon nanotube field-effect-transistor-based biosensors. Adv. Mater. **19**(11), 1439–1451 (2007)
30. I. Barkefors et al., A fluidic device to study directional angiogenesis in complex tissue and organ culture models. Lab Chip **9**(4), 529–535 (2009). https://doi.org/10.1039/B814691H
31. T. Kuila et al., Recent advances in graphene-based biosensors. Biosens. Bioelectron. **26**(12), 4637–4648 (2011)
32. K.C. Bantz et al., Recent progress in SERS biosensing. Phys. Chem. Chem. Phys. **13**(24), 11551–11567 (2011). https://doi.org/10.1039/C0CP01841D
33. B.D. Malhotra, M. Das, P.R. Solanki, Opportunities in nano-structured metal oxides based biosensors. J. Phys. Conf. Ser. **358** (2012). https://doi.org/10.1088/1742-6596/358/1/012007
34. R.W. Picard et al., Washable wearable biosensor. Google Patents, 2012
35. B. Otis et al., Wireless powered contact lens with biosensor. Google Patents, 2013
36. F. Jain et al., Implantable biosensor and methods of use thereof. Google Patents, 2014
37. A.E. Cetin et al., Handheld high-throughput plasmonic biosensor using computational on-chip imaging. Light Sci. Appl. **3**(1), e122 (2014)
38. Y. Liu et al., Surface plasmon resonance biosensor based on smart phone platforms. Sci. Rep. **5**, 12864 (2015)

39. N. Yi, M.R. Abidian, Conducting polymers and their biomedical applications, in *Biosynthetic Polymers for Medical Applications* (2016), pp. 243–276
40. S.N. Sawant, Development of biosensors from biopolymer composites, in *Biopolymer Composites in Electronics* (2017), pp. 353–383
41. D. Xu et al., Automatic smartphone-based microfluidic biosensor system at the point of care. Biosens. Bioelectron. **110**, 78–88 (2018)
42. Y. Zhao et al., Current status of optical fiber biosensor based on surface plasmon resonance. Biosens. Bioelectron. **142**, 111505 (2019). https://doi.org/10.1016/j.bios.2019.111505
43. D.V. Vokhmyanina et al., 'Artificial peroxidase' nanozyme—enzyme based lactate biosensor. Talanta **208**, 120393 (2020). https://doi.org/10.1016/j.talanta.2019.120393
44. M.F. De Volder et al., Carbon nanotubes: present and future commercial applications. Science **339**(6119), 535–539 (2013)
45. G.A. Rivas et al., Carbon nanotubes for electrochemical biosensing. Talanta **74**(3), 291–307 (2007)
46. J. Wang, Carbon-nanotube based electrochemical biosensors: a review. Electroanalysis **17**(1), 7–14 (2005)
47. H. Beitollahi et al., A review on the effects of introducing CNTs in the modification process of electrochemical sensors. Electroanalysis **31**(7), 1195–1203 (2018). https://doi.org/10.1002/elan.201800370
48. K. Maehashi, K. Matsumoto, Label-free electrical detection using carbon nanotube-based biosensors. Sensors **9**(7), 5368–5378 (2009)
49. A. Sarkar et al., Characterization of carbon nanotubes and its application in biomedical sensor for prostate cancer detection. Sens. Lett. **17**(1), 17–24 (2019)
50. H. Lilja et al., Prostate-specific antigen in serum occurs predominantly in complex with alpha 1-antichymotrypsin. Clin. Chem. **37**(9), 1618–1625 (1991)
51. S. Gomes-Filho et al., A carbon nanotube-based electrochemical immunosensor for cardiac troponin T. Microchem. J. **109**, 10–15 (2013)
52. Y. Tsujita et al., Carbon nanotube amperometric chips with pneumatic micropumps. Jpn. J. Appl. Phys. **47**(4R), 2064 (2008)
53. S.J. Tans, A.R. Verschueren, C. Dekker, Room-temperature transistor based on a single carbon nanotube. Nature **393**(6680), 49 (1998)
54. T.H. da Costa et al., A paper-based electrochemical sensor using inkjet-printed carbon nanotube electrodes. ECS J. Solid State Sci. Technol. **4**(10), S3044–S3047 (2015)
55. B. Zribi et al., A microfluidic electrochemical biosensor based on multiwall carbon nanotube/ferrocene for genomic DNA detection of *Mycobacterium tuberculosis* in clinical isolates. Biomicrofluidics **10**(1), 014115 (2016)
56. M.J. O'Connell, E.E. Eibergen, S.K. Doorn, Chiral selectivity in the charge-transfer bleaching of single-walled carbon-nanotube spectra. Nat. Mater. **4**(5), 412 (2005)
57. K. Suenaga et al., Imaging active topological defects in carbon nanotubes. Nat. Nanotechnol. **2**(6), 358 (2007)
58. P.W. Barone et al., Near-infrared optical sensors based on single-walled carbon nanotubes. Nat. Mater. **4**(1), 86 (2005)
59. E.S. Jeng et al., Detection of DNA hybridization using the near-infrared band-gap fluorescence of single-walled carbon nanotubes. Nano Lett. **6**(3), 371–375 (2006)
60. Y. Wang et al., Graphene and graphene oxide: biofunctionalization and applications in biotechnology. Trends Biotechnol. **29**(5), 205–212 (2011)
61. S. Liu, Functionalization of carbon nanomaterials for biomedical applications. C J. Carbon Res. **5**(4) (2019). https://doi.org/10.3390/c5040072
62. C.H. Lu et al., A graphene platform for sensing biomolecules. Angew. Chem. Int. Ed. **48**(26), 4785–4787 (2009)
63. Y. Shao et al., Graphene based electrochemical sensors and biosensors: a review. Electroanalysis **22**(10), 1027–1036 (2010)
64. M. Inagaki, F. Kang, Graphene derivatives: graphane, fluorographene, graphene oxide, graphyne and graphdiyne. J. Mater. Chem. A **2**(33), 13193–13206 (2014)

65. Q. He et al., Graphene-based electronic sensors. Chem. Sci. **3**(6), 1764–1772 (2012)
66. Z. Tang et al., Constraint of DNA on functionalized graphene improves its biostability and specificity. Small **6**(11), 1205–1209 (2010)
67. T. Cohen-Karni et al., Graphene and nanowire transistors for cellular interfaces and electrical recording. Nano Lett. **10**(3), 1098–1102 (2010)
68. M.A. Ali et al., Lipid–lipid interactions in aminated reduced graphene oxide interface for biosensing application. Langmuir **30**(14), 4192–4201 (2014)
69. N.G. Shang et al., Catalyst-free efficient growth, orientation and biosensing properties of multilayer graphene nanoflake films with sharp edge planes. Adv. Func. Mater. **18**(21), 3506–3514 (2008)
70. Y. Liu et al., Biocompatible graphene oxide-based glucose biosensors. Langmuir **26**(9), 6158–6160 (2010)
71. P. Zuo et al., A PDMS/paper/glass hybrid microfluidic biochip integrated with aptamer-functionalized graphene oxide nano-biosensors for one-step multiplexed pathogen detection. Lab Chip **13**(19), 3921–3928 (2013)
72. Y. Wang et al., Aptamer/graphene oxide nanocomplex for in situ molecular probing in living cells. J. Am. Chem. Soc. **132**(27), 9274–9276 (2010)
73. S. Srivastava et al., Electrophoretically deposited reduced graphene oxide platform for food toxin detection. Nanoscale **5**(7), 3043–3051 (2013)
74. A.-M.J. Haque et al., An electrochemically reduced graphene oxide-based electrochemical immunosensing platform for ultrasensitive antigen detection. Anal. Chem. **84**(4), 1871–1878 (2012)
75. M.A. Ali et al., Mesoporous few-layer graphene platform for affinity biosensing application. ACS Appl. Mater. Interfaces **8**(12), 7646–7656 (2016)
76. C. Zhu et al., Electrochemical sensors and biosensors based on nanomaterials and nanostructures. Anal. Chem. **87**(1), 230–249 (2014)
77. S.H. Hu et al., Quantum-dot-tagged reduced graphene oxide nanocomposites for bright fluorescence bioimaging and photothermal therapy monitored in situ. Adv. Mater. **24**(13), 1748–1754 (2012)
78. V. Vamvakaki, K. Tsagaraki, N. Chaniotakis, Carbon nanofiber-based glucose biosensor. Anal. Chem. **78**(15), 5538–5542 (2006)
79. V. Vamvakaki, M. Hatzimarinaki, N. Chaniotakis, Biomimetically synthesized silica–carbon nanofiber architectures for the development of highly stable electrochemical biosensor systems. Anal. Chem. **80**(15), 5970–5975 (2008)
80. R.K. Gupta et al., Label-free detection of C-reactive protein using a carbon nanofiber based biosensor. Biosens. Bioelectron. **59**, 112–119 (2014)
81. P.U. Arumugam et al., Wafer-scale fabrication of patterned carbon nanofiber nanoelectrode arrays: a route for development of multiplexed, ultrasensitive disposable biosensors. Biosens. Bioelectron. **24**(9), 2818–2824 (2009)
82. M. Rahman et al., A comprehensive review of glucose biosensors based on nanostructured metal-oxides. Sensors **10**(5), 4855–4886 (2010)
83. P.R. Solanki et al., Nanostructured metal oxide-based biosensors. NPG Asia Mater. **3**(1), 17 (2011)
84. F. Xiao et al., Growth of metal–metal oxide nanostructures on freestanding graphene paper for flexible biosensors. Adv. Func. Mater. **22**(12), 2487–2494 (2012)
85. J.M. Pingarrón, P. Yanez-Sedeno, A. González-Cortés, Gold nanoparticle-based electrochemical biosensors. Electrochim. Acta **53**(19), 5848–5866 (2008)
86. Y.K. Yoo et al., Gold nanoparticles assisted sensitivity improvement of interdigitated microelectrodes biosensor for amyloid-β detection in plasma sample. Sens. Actuators B Chem. 127710 (2020)
87. J. Lin et al., One-step synthesis of silver nanoparticles/carbon nanotubes/chitosan film and its application in glucose biosensor. Sens. Actuators B Chem. **137**(2), 768–773 (2009)
88. J. Chen et al., A fluorescent biosensor based on catalytic activity of platinum nanoparticles for freshness evaluation of aquatic products. Food Chem. **310**, 125922 (2020)

89. M. Balaji et al., Fabrication of palladium nanoparticles anchored polypyrrole functionalized reduced graphene oxide nanocomposite for antibiofilm associated orthopedic tissue engineering. Appl. Surf. Sci. 145403 (2020)
90. A.J. Haes et al., A nanoscale optical biosensor: the long range distance dependence of the localized surface plasmon resonance of noble metal nanoparticles. J. Phys. Chem. B **108**(1), 109–116 (2004)
91. G. Doria et al., Noble metal nanoparticles for biosensing applications. Sensors **12**(2), 1657–1687 (2012)
92. S. Scarano et al., Surface plasmon resonance imaging for affinity-based biosensors. Biosens. Bioelectron. **25**(5), 957–966 (2010)
93. J. Wang, Electrochemical biosensing based on noble metal nanoparticles. Microchim. Acta **177**(3–4), 245–270 (2012)
94. K.-S. Lee, M.A. El-Sayed, Gold and silver nanoparticles in sensing and imaging: sensitivity of plasmon response to size, shape, and metal composition. J. Phys. Chem. B **110**(39), 19220–19225 (2006)
95. C. Sönnichsen et al., A molecular ruler based on plasmon coupling of single gold and silver nanoparticles. Nat. Biotechnol. **23**(6), 741 (2005)
96. S. Zhang et al., Covalent attachment of glucose oxidase to an Au electrode modified with gold nanoparticles for use as glucose biosensor. Bioelectrochemistry **67**(1), 15–22 (2005)
97. D.J. Maxwell, J.R. Taylor, S. Nie, Self-assembled nanoparticle probes for recognition and detection of biomolecules. J. Am. Chem. Soc. **124**(32), 9606–9612 (2002)
98. S. Hassanpour, A. Saadati, M. Hasanzadeh, pDNA conjugated with citrate capped silver nanoparticles towards ultrasensitive bio-assay of haemophilus influenza in human biofluids: a novel optical biosensor. J. Pharm. Biomed. Anal. **180**, 113050 (2020)
99. A.J. Haes, R.P. Van Duyne, A nanoscale optical biosensor: sensitivity and selectivity of an approach based on the localized surface plasmon resonance spectroscopy of triangular silver nanoparticles. J. Am. Chem. Soc. **124**(35), 10596–10604 (2002)
100. J.-D. Qiu, S.-G. Cui, R.-P. Liang, Hydrogen peroxide biosensor based on the direct electrochemistry of myoglobin immobilized on ceria nanoparticles coated with multiwalled carbon nanotubes by a hydrothermal synthetic method. Microchim. Acta **171**(3–4), 333–339 (2010)
101. N. Cheng et al., Amperometric glucose biosensor based on integration of glucose oxidase with palladium nanoparticles/reduced graphene oxide nanocomposite. Am. J. Anal. Chem. **3**(04), 312 (2012)
102. M. Yang et al., Layer-by-layer self-assembled multilayer films of carbon nanotubes and platinum nanoparticles with polyelectrolyte for the fabrication of biosensors. Biomaterials **27**(2), 246–255 (2006)
103. A. Kaushik et al., Organic–inorganic hybrid nanocomposite-based gas sensors for environmental monitoring. Chem. Rev. **115**(11), 4571–4606 (2015)
104. M. Vaseem, A. Umar, Y.-B. Hahn, ZnO nanoparticles: growth, properties, and applications, in *Metal Oxide Nanostructures and Their Applications*, vol. 5 (2010), pp. 1–36
105. X. Lu et al., Porous nanosheet-based ZnO microspheres for the construction of direct electrochemical biosensors. Biosens. Bioelectron. **24**(1), 93–98 (2008)
106. A. Lazarides et al., Optical properties of metal nanoparticles and nanoparticle aggregates important in biosensors. J. Mol. Struct. (Thoechem) **529**(1–3), 59–63 (2000)
107. N. Dimcheva, Nanostructures of noble metals as functional materials in biosensors. Curr. Opin. Electrochem. **19**, 35–41 (2020). https://doi.org/10.1016/j.coelec.2019.09.008
108. A. Kawamura, T. Miyata, Biosensors, in *Biomaterials Nanoarchitectonics* (2016), pp. 157–176
109. M.-A. Neouze, U. Schubert, Surface modification and functionalization of metal and metal oxide nanoparticles by organic ligands. Monatsh. Chem. Chem. Mon. **139**(3), 183–195 (2008)
110. M. Schulz-Dobrick, K.V. Sarathy, M. Jansen, Surfactant-free synthesis and functionalization of gold nanoparticles. J. Am. Chem. Soc. **127**(37), 12816–12817 (2005)
111. A. Ulman, Formation and structure of self-assembled monolayers. Chem. Rev. **96**(4), 1533–1554 (1996)

112. R.C. Doty et al., Extremely stable water-soluble Ag nanoparticles. Chem. Mater. **17**(18), 4630–4635 (2005)
113. E. Ramirez et al., Influence of organic ligands on the stabilization of palladium nanoparticles. J. Organomet. Chem. **689**(24), 4601–4610 (2004)
114. C. Lynch, An overview of first-principles calculations of NMR parameters for paramagnetic materials. Mater. Sci. Technol. **32**(2), 181–194 (2016)
115. J. Liu, Y. Lu, Fast colorimetric sensing of adenosine and cocaine based on a general sensor design involving aptamers and nanoparticles. Angew. Chem. Int. Ed. **45**(1), 90–94 (2006)
116. M.A. Ali et al., Protein-conjugated quantum dots interface: binding kinetics and label-free lipid detection. Anal. Chem. **86**(3), 1710–1718 (2014)
117. F. Liu et al., Surface characterization study on SnO_2 powder modified by thiourea. Mater. Chem. Phys. **93**(2–3), 301–304 (2005)
118. Q.-L. Zhang et al., Particle-size-dependent distribution of carboxylate adsorption sites on TiO_2 nanoparticle surfaces: insights into the surface modification of nanostructured TiO_2 electrodes. J. Phys. Chem. B **108**(39), 15077–15083 (2004)
119. S. Yu, G.M. Chow, Carboxyl group (–CO_2H) functionalized ferrimagnetic iron oxide nanoparticles for potential bio-applications. J. Mater. Chem. **14**(18), 2781–2786 (2004)
120. R.M. Petoral Jr. et al., Organosilane-functionalized wide band gap semiconductor surfaces. Appl. Phys. Lett. **90**(22), 223904 (2007)
121. C.D. Sanchez et al., Designed hybrid organic–inorganic nanocomposites from functional nanobuilding blocks. Chem. Mater. **13**(10), 3061–3083 (2001)
122. C. Sanchez et al., Applications of hybrid organic–inorganic nanocomposites. J. Mater. Chem. **15**(35–36), 3559–3592 (2005)
123. P.M. Ajayan, L.S. Schadler, P.V. Braun, *Nanocomposite Science and Technology* (Wiley, 2006)
124. Y. Xian et al., Glucose biosensor based on Au nanoparticles–conductive polyaniline nanocomposite. Biosens. Bioelectron. **21**(10), 1996–2000 (2006)
125. X.-L. Luo et al., Electrochemically deposited nanocomposite of chitosan and carbon nanotubes for biosensor application. Chem. Commun. **16**, 2169–2171 (2005)
126. X. Ma et al., A functionalized graphene oxide-iron oxide nanocomposite for magnetically targeted drug delivery, photothermal therapy, and magnetic resonance imaging. Nano Res. **5**(3), 199–212 (2012)
127. F. Croce et al., Nanocomposite polymer electrolytes for lithium batteries. Nature **394**(6692), 456 (1998)
128. K.H. An et al., Enhanced sensitivity of a gas sensor incorporating single-walled carbon nanotube–polypyrrole nanocomposites. Adv. Mater. **16**(12), 1005–1009 (2004)
129. Q. Zhang et al., High refractive index inorganic–organic interpenetrating polymer network (IPN) hydrogel nanocomposite toward artificial cornea implants. ACS Macro Lett. **1**(7), 876–881 (2012)
130. C. Wang et al., Gold nanoclusters and graphene nanocomposites for drug delivery and imaging of cancer cells. Angew. Chem. Int. Ed. **50**(49), 11644–11648 (2011)
131. C. Barbas Arribas, D. Rojo Blanco, Understanding the antimicrobial mechanism of TiO_2-based nanocomposite films in a pathogenic Bacterium/David Rojo … [et al.]. (2015)
132. T. Ahuja, D. Kumar, Recent progress in the development of nano-structured conducting polymers/nanocomposites for sensor applications. Sens. Actuators B Chem. **136**(1), 275–286 (2009)
133. P. Visakh, C. Della Pina, E. Falletta, *Polyaniline Blends, Composites, and Nanocomposites* (Elsevier, 2017)
134. K. Rege et al., Enzyme–polymer–single walled carbon nanotube composites as biocatalytic films. Nano Lett. **3**(6), 829–832 (2003)
135. X. Kang et al., Glucose oxidase–graphene–chitosan modified electrode for direct electrochemistry and glucose sensing. Biosens. Bioelectron. **25**(4), 901–905 (2009)
136. X.F. Wang et al., Signal-on electrochemiluminescence biosensors based on CdS–carbon nanotube nanocomposite for the sensitive detection of choline and acetylcholine. Adv. Func. Mater. **19**(9), 1444–1450 (2009)

137. S. Wang et al., Amperometric tyrosinase biosensor based on Fe_3O_4 nanoparticles–chitosan nanocomposite. Biosens. Bioelectron. **23**(12), 1781–1787 (2008)
138. S. Kailasa et al., NiO nanoparticles-decorated conductive polyaniline nanosheets for amperometric glucose biosensor. Mater. Chem. Phys. **242**, 122524 (2020)
139. Q. Yao et al., Enhanced thermoelectric performance of single-walled carbon nanotubes/polyaniline hybrid nanocomposites. ACS Nano **4**(4), 2445–2451 (2010)
140. H. Zhang et al., Fe_3O_4/polypyrrole/Au nanocomposites with core/shell/shell structure: synthesis, characterization, and their electrochemical properties. Langmuir **24**(23), 13748–13752 (2008)
141. N. Ruecha et al., Novel paper-based cholesterol biosensor using graphene/polyvinylpyrrolidone/polyaniline nanocomposite. Biosens. Bioelectron. **52**, 13–19 (2014)
142. F. Croisier, C. Jérôme, Chitosan-based biomaterials for tissue engineering. Eur. Polym. J. **49**(4), 780–792 (2013)
143. A. Kaushik et al., Chitosan–iron oxide nanobiocomposite based immunosensor for ochratoxin-A. Electrochem. Commun. **10**(9), 1364–1368 (2008)
144. R. Khan et al., Zinc oxide nanoparticles-chitosan composite film for cholesterol biosensor. Anal. Chim. Acta **616**(2), 207–213 (2008)
145. W.-D. Zhang, B. Xu, L.-C. Jiang, Functional hybrid materials based on carbon nanotubes and metal oxides. J. Mater. Chem. **20**(31), 6383–6391 (2010)
146. J. Chen et al., Solution properties of single-walled carbon nanotubes. Science **282**(5386), 95–98 (1998)
147. G.-X. Wang et al., Manganese oxide/MWNTs composite electrodes for supercapacitors. Solid State Ionics **176**(11–12), 1169–1174 (2005)
148. B. Liu et al., Preparation of Pt/MgO/CNT hybrid catalysts and their electrocatalytic properties for ethanol electrooxidation. Energy Fuels **21**(3), 1365–1369 (2007)
149. A. Kongkanand, P.V. Kamat, Electron storage in single wall carbon nanotubes. Fermi level equilibration in semiconductor–SWCNT suspensions. ACS Nano **1**(1), 13–21 (2007)
150. T. Sainsbury, D. Fitzmaurice, Templated assembly of semiconductor and insulator nanoparticles at the surface of covalently modified multiwalled carbon nanotubes. Chem. Mater. **16**(19), 3780–3790 (2004)
151. N. Chauhan, C.S. Pundir, An amperometric biosensor based on acetylcholinesterase immobilized onto iron oxide nanoparticles/multi-walled carbon nanotubes modified gold electrode for measurement of organophosphorus insecticides. Anal. Chim. Acta **701**(1), 66–74 (2011)
152. M. Das et al., Zirconia grafted carbon nanotubes based biosensor for *M. tuberculosis* detection. Appl. Phys. Lett. **99**(14), 143702 (2011)
153. S.J. Aravind, S. Ramaprabhu, Dopamine biosensor with metal oxide nanoparticles decorated multi-walled carbon nanotubes. Nanosci. Methods **1**(1), 102–114 (2012)
154. Q. Shen et al., Electrochemical biosensing for cancer cells based on TiO_2/CNT nanocomposites modified electrodes. Electroanalysis **20**(23), 2526–2530 (2008)
155. S. Hrapovic et al., Electrochemical biosensing platforms using platinum nanoparticles and carbon nanotubes. Anal. Chem. **76**(4), 1083–1088 (2004)
156. X.-C. Dong et al., 3D graphene–cobalt oxide electrode for high-performance supercapacitor and enzymeless glucose detection. ACS Nano **6**(4), 3206–3213 (2012)
157. K. Yang et al., Nano-graphene in biomedicine: theranostic applications. Chem. Soc. Rev. **42**(2), 530–547 (2013)
158. Z. Wang et al., Direct electrochemical reduction of single-layer graphene oxide and subsequent functionalization with glucose oxidase. J. Phys. Chem. C **113**(32), 14071–14075 (2009)
159. Y. Zhao et al., ZnO-nanorods/graphene heterostructure: a direct electron transfer glucose biosensor. Sci. Rep. **6**, 32327 (2016)

Bionanomaterial Thin Film for Piezoelectric Applications

Mohd Hatta Maziati Akmal and Farah Binti Ahmad

Abstract Wearable and flexible electronics are recently gaining interest due to the expansion of Internet of things (IoT). Thin-film piezoelectric materials have the potential to be used in the development of flexible electronic devices in energy harvesting, sensing, and biomedicine. This is mainly because of the inherent ability of piezoelectric materials to convert the mechanical energy into the electrical energy and vice versa. Piezoelectricity in material represents the property of certain crystalline structures that are capable of developing electricity when pressure is applied. However, conventional piezoelectric materials such as PZT (lead zirconate titanate) and PVDF (poly(vinylidene flouride)) are expensive, non-renewable, non-biodegradable, and lack biocompatibility due to the cytotoxic nature of lead-based materials. Piezoelectric materials from natural polymers of biomaterials may provide a solution for the drawbacks of piezoceramics and piezoelectric polymers. This review's emphasis is on the piezoelectricity of various bionanomaterials (cellulose, chitin, chitosan, collagen, amino acid, and peptide). The various methods used to measure piezoelectricity of biomaterials are also discussed. This study shows that biomaterials have the potential to be used as piezoelectric nanogenerators in energy harvesting, sensors and biosensors, as well as in cell and tissue engineering, wound healing and drug delivery.

Keywords Bionanomaterial · Piezoelectric · Nanocrystalline · Cellulose · Chitin · Chitosan · Collagen · Peptide

---

M. H. M. Akmal
Department of Science in Engineering, Faculty of Engineering, International Islamic University Malaysia, Kuala Lumpur, Malaysia

F. B. Ahmad (✉)
Department of Biotechnology Engineering, Faculty of Engineering, International Islamic University Malaysia, Kuala Lumpur, Malaysia
e-mail: farahahmad@iium.edu.my

A. T. Jameel and A. Z. Yaser (eds.), *Advances in Nanotechnology and Its Applications*,
https://doi.org/10.1007/978-981-15-4742-3_4

1 Introduction

Piezoelectric materials can transduce the mechanical pressure acting on them to electrical signals (direct piezoelectric effect) and electrical signals to mechanical signals (converse piezoelectric effect) [1]. Therefore, piezoelectric materials have vast potential in the application of energy harvesters, biomedical devices, and sensors. When the piezoelectric element is deformed, electric polarization occurs in proportion to the pressure, in which a potential difference occurs until the space charge corrects this polarization [2]. The fundamental principle of piezoelectric effect is attributed to the separation of the center of neutrality of charges on the crystal lattice as the material is deformed along certain axes [3]. The piezoelectricity of materials depends on its crystal lattice structure and the lack of a center of symmetry [1]. Typically, material that exhibits piezoelectricity consists of anisotropic crystals which are the crystals that void the center of symmetry [4].

Piezoelectric materials can also be classified as piezoceramic and piezoelectric polymers. Among piezoceramic materials are lead zirconate titanate (PZT), aluminum nitride (AlN), aluminum phosphate (berlinite), barium titanate ($BaTiO_3$), crystalized topaz ($Al_2SiO_4(F,OH)_2$), gallium orthophosphate, lithium niobate ($LiNbO_3$), quartz (SiO_2), tartrate tetrahydrate (Rochelle salt), and zinc oxide (ZnO) [4, 5]. Piezoceramics such as PZT can typically generate a very high piezoelectric coefficient (from 200 to 350 pC/N); however, they are non-renewable and not biodegradable [3]. PZT is not suitable for biomedical application as it may pose a toxicity risk to human cells [6]. PVDF (poly(vinylidene fluoride)) is a well-known piezoelectric copolymer in biomedical application due to its significant piezoelectric coefficient of 20 pC/N, its flexibility and non-toxicity [3]. However, PVDF is a non-biodegradable polymer. Fluorine-based electro-active polymers may also release toxic gases such as hydrogen fluoride into the environment during their syntheses and decomposition processes [6]. Current commercially available flexible piezoelectric films made of PVDF are also relatively expensive [7].

Bionanomaterials are nanocrystallites which could be retrieved from natural resources or it can be made in the laboratory such as chitosan, collagen, elastin, and natural biomaterials such as fish scale. Bionanomaterials which are biodegradable mostly used in the biomedical field for replacing tissues, organs or any functions in the human body. Energy harvesting from the bionanomaterials has become an interesting topic in which a "self-powered device" is being used instead of using rechargeable or electric powered devices. These "self-powered" devices are made from the vibrations of the particles inside the materials itself as it is a requirement for the sustainable and independent operation of less power-consuming system. Piezoelectric biomaterials can be made into a thin film, which has the potential to be used as stretchable, flexible, non-bulky, slim, self-powered, and cheap wearable electronic devices. Piezoelectric bionanomaterials have potential application in (1) nanogenerators (energy harvesters), (2) actuators, sensors and biosensors, and (3) biomedicine which includes cell and tissue engineering, wound healing and drug delivery.

Piezoelectric bionanomaterials can be used as smart devices in biomedical applications due to its flexibility as thin film, biodegradability, and biocompatibility. Piezoelectric materials may provide the solution to bulky systems that are conventionally applied in biomedical devices. The bulky systems are not suitable as implantable medical devices inside the body due to incompatible contact with the curved and corrugated surfaces of human organs such as the heart, blood vessel, eye, brain, and lungs [8]. Flexible piezoelectric bionanomaterial thin films may provide alternatives to the existing electronic health-monitoring devices involving heart rate, wrist pulse motion, blood pressure, intraocular pressure monitoring. Piezoelectric bionanomaterial thin film is of current interest; unlike conventional health-monitoring systems (e.g., blood pressure meters), it can be made into portable and wearable health-monitoring devices that may provide real-time, continuous, recorded data related to health conditions [9].

The expansion of smart devices generation on the Internet of things (IoT) has raised concerns about the associated waste and emission generation [7]. Using biodegradable piezoelectric membranes for local energy production could enhance the IoT in an environmentally and economically sound fashion compared to non-degradable materials or batteries [7]. Piezoelectric material could be derived from biomaterial that may provide the solution for the toxicity and non-biodegradability issues of the conventional piezoelectric materials.

As biomaterials are derived from natural resources, it is typically non-toxic. Therefore, the biomaterials are biocompatible and may broaden piezoelectric application to medical devices. The biocompatibility of piezoelectric materials is especially essential for the applications of implantable piezoelectric devices in the human or animal body. The use of biomaterials may enhance the sustainability of piezoelectric application due to its origin and biodegradability. Most biomaterials can be derived from waste products, such as chitin or chitosan from crab, prawn shells and squid pen, cellulose from onion skins, and collagen from fish scale and fish swim bladder [6, 10–12]. This review aims to assess the potential use of bionanomaterials thin film in the application of piezoelectric nanogenerator, sensors and biosensors, and cell and tissue engineering. This review also discusses various methods on how to quantify piezoelectricity on bionanomaterials.

2 The Mechanism Behind Piezoelectricity

The existence of piezoelectricity is due to the reorientation of molecular dipoles within the intrinsic or crystalline structure of materials [13]. The piezoelectricity, thus, can be achieved via the application of high electric field or mechanical force in different directions on bulk materials [14, 15]. Therefore, the piezoelectric effect in bionanomaterials can be manifested in two ways:

1. The direct effect, when the electrical charge is generated as a result of mechanical stress applied to the crystal and;

2. The converse effect, when a strain is generated as a result of an electrical field applied to crystals.

The constitutive equations related to these two effects are described by two basic equations that relate the elastic variables, stress (T) and strain (S) to the electric variables, electric field (E), and dielectric displacement (D). Dielectric represents the electric flux density per unit area, having the properties of a vector while stress and strain are tensors. The two equations of state are written as follows:

$$D = dT + \varepsilon^T E \tag{1}$$

$$S = dE + s^E T \tag{2}$$

The first equation describes the direct effect meanwhile the second equation is the converse effect. Here, d is a piezoelectric coefficient known generally as a charge constant and expressed as 10^{-2} pC/N for direct effect and 10^{-2} m/V for converse effect.

The central idea of how a piezoelectric material functions is shown in Fig. 1. Basically, the charges in a piezoelectric material are exactly balanced (see Fig. 1a). At this stage, the effects of the charges exactly cancel out with no net charge on the faces of the crystal (Fig. 1b). Conversely, if the material was compressed, the material drives the charges to imbalance (Fig. 1c). As a result, the charges dipole moments do not cancel each other, and net negative and positive charges emerge on the opposite faces of the crystal. By compressing the crystal, the voltage was generated across its opposed faces, thus producing piezoelectricity (Fig. 1d).

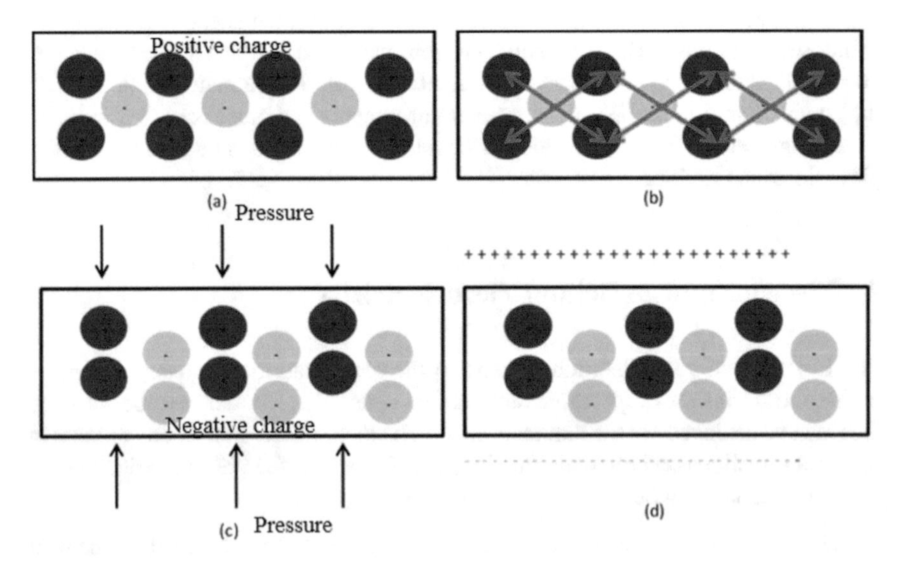

Fig. 1 Schematic of piezoelectric working principle

A center of symmetry is a place where the crystal structure displays inversion symmetry. It can also be said that if the crystal has an atom at (x, y, z) with reference to the center of symmetry, it must also have an atom at $(-x, -y, -z)$. In essence, materials that possess no center symmetry in their crystalline structure possess piezoelectricity [16, 17]. Conversely, materials that are centrosymmetric when put under stress are subjected to the symmetrical shift of orientation of the ions that result in zero net polarization.

Essentially, the piezoelectricity can be quantified by the electromechanical coupling factor, k. The high value for k is preferable in piezoelectric materials because it indicates the accuracy of energy conversion [18, 19]. The amount of k is generally less than one because the energy transformation cannot be perfect. Apart from that, the mechanical factor or quality factor, Q_{m}, is crucial to measure the external contribution to improve piezoelectric response. Principally, the high-quality factor, Q_{m}, is attributed to the improvement in domain orientation which causes the enhancement of piezoelectric phenomenon within the crystal structure [20].

Another significant requirement for the material to be regarded as good piezoelectric material is the high value of dielectric permittivity, ε, and the low dissipation factor or dielectric loss, tan δ. The permittivity is significant to measure a charge gathered on the piezoelectric active region; meanwhile, dielectric loss governs the loss rate energy of the electromechanical system. As per the Institute of Electrical and Electronics Engineers (IEEE) convention, the constitutive equation associated with piezoelectric properties can be represented as follows (*An American National Standard IEEE Standard on Piezoelectricity*, 1988):

$$k = \sqrt{\frac{\pi}{2}\frac{f_{\mathrm{r}}}{f_{\mathrm{a}}}\tan\left(\frac{f_{\mathrm{a}} - f_{\mathrm{r}}}{f_{\mathrm{a}}}\right)} \tag{3}$$

In Eq. (3), f_{r} (Hz) is the resonance frequency, while f_{a} (Hz) represents antiresonance frequency. For the quality factor, Q_{m}, the constitutive equation is as follows:

$$Q_{\mathrm{m}} = \frac{1}{R}\sqrt{\frac{L}{C}} \tag{4}$$

where R (Ω), L (H), and C (F/m) in Eq. (4) can be defined as the resistance, inductance, and capacitance in the equivalent electrical circuit, respectively.

When the piezoelectric material is placed between two plates of capacitor, its capacitance is increased by a factor ε. This parameter is also called dielectric permittivity as shown in Eq. (5). Mathematically, the permittivity (ε) can be represented as;

$$\varepsilon = \frac{Ct}{\varepsilon_0 A} \tag{5}$$

where t (m) is the distance of electrode within piezoelectric, A (m^2) is an active area, and ε_0 is the permittivity of free space (8.85×10^{-12} F/m). The loss energy rate denoted by dielectric loss ($tan\ \delta$) can be obtained by the ratio of imaginary part of permittivity (ε'') to real part of permittivity (ε').

3 Bionanomaterials for Piezoelectric Application

Various biological materials, such as wood, bone, and fibrous proteins like collagen and chitin, have been discovered for their piezoelectric properties for so many years ago [13]. Most biomaterials that exhibit piezoelectricity are natural polymers which can be categorized as amino acid, cellulose, chitin, chitosan, and collagen.

3.1 Cellulose

Cellulose is the most abundant natural polymer on Earth that can be found in most plant-based materials such as wood and lignocellulosic materials, from agricultural biomass residues [21]. Cellulose is a polysaccharide that is made of repeating units of β-1,4-linked D-glucose monomer units. Nanocellulose material is becoming a growing interest attributed to its excellent inherent physical and chemical properties which include high tensile strength and elastic modulus (130–150 GPa), high specific surface area, low density (1.6 g/cm^3), reactive surfaces, biodegradable, and renewable [22]. There are two types of nanocellulose that can be extracted from natural resources, which are cellulose nanocrystals (CNC) (i.e., cellulose whiskers), cellulose nanofibrils (CNF) (i.e., nanofibrillated cellulose (NFC) or microfibrillated cellulose (MFC)), and bacteria nanocellulose [23]. The extraction of CNC involved chemical hydrolysis of the amorphous regions of cellulose, whereas CNF can be produced by subjecting a cellulose suspension to a mechanical process [23]. As piezoelectricity was discovered in wood a long time ago, it is expected that cellulose can exhibit the piezoelectric properties [24]. As piezoelectric effect occurs due to polarization density change and dipole moments within a material, a highly crystalline structure will exhibit significant piezoelectricity [25]. The piezoelectricity of crystalline nanoscale materials, such as CNC and CNF, is projected to be more prominent than in wood [26]. The orientation of CNC crystals and its permanent dipole moment have a major effect on the overall piezoelectricity [27]. The piezoelectric properties of cellulose could also be attributed to its anisotropic triclinic and monoclinic unit crystal structure that is related to unevenly distributed carbon atoms and change of polarization density of charged atomic groups under electric fields [25].

Among cellulose that has been tested for piezoelectric properties was bacterial nanocellulose, CNF and CNC from bleached birch cellulose mass and onion skin [24, 26–28]. Amorphous cellulose chains and crystalline CNC regions are the constituents of CNF [26], which have resulted in a substantial piezoelectric affect. The piezoelectricity effect in CNF film could be optimized through the polarization of the film that may result into the orientation of crystalline CNC regions inside the CNF film [26]. Even though the highly crystalline structure of CNC renders it promising for piezoelectric application, it has some drawbacks in the making of thin film. The preparation of thin film from CNC via solvent casting from the highly fluid dispersion was not feasible [7]. Bacterial nanocellulose was reported to possess higher piezoelectricity than wood at 5–20 pC/N [28]. The main composition of bacteria nanocellulose is α-cellulose, which allows ease of modification on the nanocellulose [23]. Ribbon-like nanofibrils can be obtained from bacteria nanocellulose naturally without mechanical treatment [23].

3.2 Chitin

Chitin is the second most abundant natural polymer after cellulose. It is a polysaccharide of poly-β-(1,4)-*N*-acetyl-*D*-glucosamine. Chitin can be found in nature from the exoskeleton of arthropods (e.g., crab, prawn), the endoskeleton of cephalopods (e.g., squid), and the cell wall of fungi and mushroom [29]. Similarly to cellulose, chitin occurs naturally as supramolecular crystalline nanofibers (3–5 nm) [29]. There are two polymorphic forms of chitin which are α and β. α-chitin is the most abundant and stable polymorph [30], in which α-chitin can be found naturally in the shells of crustaceans, the cuticle of insects and the cell wall of fungi, whereas β-chitin is found in squid pen [31]. Chitin is insoluble and recalcitrant in nature due to strong hydrogen bonds that hold the nanofibers to form fiber bundles [29]. Chitin is soluble in high concentration of inorganic acids and other solvents (e.g., dichloro- and trichloroacetic acids, $CaBr_2{\cdot}H_2O$ saturated methanol, hexafluoroacetone, hexafluoroisopropyl alcohol, lithium thiocyanate, *N*-methyl-2-pyrrolidone, and *N*,*N*-dimethylacetamide DMA, and LiCl mixture) that are mostly, corrosive, dangerous, and toxic [30].

Most studies on the piezoelectricity of chitin were on chitin that were sourced from marine organisms, including crab shell and prawn shell. Hoque et al. investigated the piezoelectricity of chitin nano-fiber (ChNF) thin film and ChNF/PVDF composite thin film, where chitin was sourced from crab shells [10]. The study showed that ChNF/PVDF composite has superior piezoelectric sensitivity (35.56 pC/N), dielectric properties, charge storage ability than stand-alone ChNF thin film [10]. This could be attributed to the robust interaction between ChNF and PVDF through synergetic effect of H-bond and uniform distribution of both materials in matrix [10]. This paper did not discuss any mechanical (e.g., tensile strength) or biological (e.g., biodegradability) properties improvement of pure ChNF thin film relative to the composite film. The crystallinity of chitin can be improved through the electrospinning method. Electrospun ChNF was reported to have a higher piezoelectric sensitivity than native ChNF due to the aligned crystalline structure of electrospun ChNF [32].

3.3 Chitosan

Chitosan (*β*-(1,4)-*D*-glucosamine) can be derived from chitin through the deacetylation process. The removal of the acetyl functional group from the chitin structure enables various enhancements of biological, chemical, and mechanical properties of chitosan. The *D*-glucosamine group of chitosan, composed of a free amino group, can be protonated that leads to solubility in weak acids and most organic solvents, and the antimicrobial properties of chitosan [30].

There are very limited studies on the use of chitosan for piezoelectric applications. Most studies utilized commercial chitosan as the source of chitosan thin film. Praveen et al. demonstrated the presence of non-centrosymmetry of chitosan thin films that were prepared using four different weak acids. The study revealed that thin film prepared using formic acid resulted in the highest second harmonic generation (SHG) intensity and the most flexible thin film in comparison with those prepared in acetic, adipic, and succinic acids [33]. A study by Hanninen et al. compared the piezoelectric performance of chitosan and chitosan/CNC composite thin films [34]. Pure chitosan thin film exhibited higher piezoelectricity sensitivity in comparison with the composite thin film at 25 pC/N [34]. Composite films showed a low piezoelectric sensitivity value possibly due to poor structure organization in composite film [7]. The addition of chitosan into the CNF composite has improved the flexibility and tensile strength of the thin film. A recent development on piezoelectricity of chitosan extracted from fungal biomass showed the potential of using sustainable source of chitosan for the production of piezoelectric chitosan thin film [35].

3.4 Collagen

Collagen is the most abundant protein in the animal kingdom [36]. Collagen molecules are rod-like triple helices, which are 300 nm in length and 1.5 nm in diameter [36]. The constituents of collagen are three polypeptide chains (α-chain) with two identical α1(I) chains and a different α2(I) chain [11]. Each α-chain is typically consisted of a repeated triplet amino acid sequence of glycine–proline–hydroxyproline [11]. The piezoelectricity of collagen could be as the result of the intermolecular hydrogen bond formation in collagen chains that generate uniaxial orientation of molecular dipoles [24]. The compact alignment of highly ordered α-helices and their inherent polarization gives rise to piezoelectricity [37]. The helical structure repeatedly aligns the dipoles of the backbone amino acids and leads to a significant permanent polarization [37]. Collagen fibers have a high degree of axial alignment of collagen molecules [36].

Most reports on the use of collagen for piezoelectric application utilized waste products that could be exploited directly prior to any chemical pretreatment such as eggshell membrane, fish bladder, fish scale, and fish skin [11, 12, 38, 39]. The piezoelectricity of eggshell membrane, that is composed of various types of highly collagenized fibrous (types I, V, and X collagen) and other different proteins including osteoprotein, sialoprotein and keratin, was studied [38]. The study discovered combined piezoelectric properties of existing collagen and other proteins [38]. The electrical dipole moment in the eggshell membrane lattice crystal could be the result of the deformation of the triple helical structure [38]. The study of fish scale for piezoelectric application deducted that the piezoelectricity and spontaneous polarization originated from molecular dipoles that rose due to the intrinsic presence of polar uniaxial orientation of hydrogen bonding motifs between the polypeptide chains [11].

3.5 Amino Acid and Peptide

The main amino acids, except for glycine, are chiral that displays the piezoelectric effect [40]. Amino acids crystallize primarily in low symmetry orthorhombic (L-try, L-threonine, L-alanine, L-serine, L-proline, L-glutamine, L-glutamate, glycine) and monoclinic (L-leucine, L-lysine, L-isoleucine, L-arginine, L-aspartate, L-cysteine, L-methionine, L-valine, L-histidine, L-asparagine, L-phenylalanine) space groups which are non-centrosymmetric [13].

One of the peptides that have been extensively studied for piezoelectricity is diphenylalanine (FF) peptide, which is a short dipeptide composed of two phenylalanine through amide-bond formation [41]. FF peptide possesses the ability of self-assembly, in which the hexagonal structure of self-assembled FF peptide nanotubes provides excellent mechanical and piezoelectric properties [42]. The application of control of polarization with an electric field during the peptide self-assembly process resulted in enhanced piezoelectric constant d_{33} at 17.9 pm/V [43].

3.6 Optimizing Piezoelectric Properties in Bionanomaterials

Piezoelectric biomaterials typically have low piezoelectric constant relative to conventional piezoelectric materials such as PZT. However, the degree of crystallinity and the molecular orientation could be engineered in order to enhance its piezoelectric properties [5]. The piezoelectricity effect of bionanomaterials can also be optimized through polarization, crystalline nanostructure alignment, film fabrication method, and moisture content reduction [25, 27]. The molecular alignment of nanomaterial can be optimized for piezoelectricity through the poling procedure. The poling procedure involves applying an electric field on a processed film for a certain period of time in order to generate piezoelectric properties [26].

4 State of the Art in Quantifying the Piezoelectricity of Bionanomaterials

The ability to quantify the piezoelectricity of biomaterials at the nanoscale is the key for the development of piezoelectric bionanomaterials. The rapid advances in piezoelectric thin-film bionanomaterials are closely related to the discovery and the improvement of the instruments used to quantify the piezoelectricity. To thoroughly grasp the behavior of piezoelectric nanomaterials, the intrinsic structure such as structural shifts and crystallinity should not be overlooked. This allows researchers to discover the constitutive relations that define the piezoelectricity in bionanomaterials.

4.1 Quasi-static Method

The quasi-static method is a direct technique which collects the piezoelectric charge value from the piezoelectric material sandwiched between the top and bottom electrodes (Fig. 2). When a uniaxial load is applied to the sample, the electric charge, *Q*, is generated due to the mechanical action. Two possible modes can be produced during charge collection which are d_{31} and d_{33} modes. For the d_{33} mode, the generated voltage is in the same direction as that of the applied stress. This is depicted in Fig. 3a. On the other hand, the d_{31} mode produces a voltage that is in a direction that is perpendicular to the stress that has been applied (Fig. 3b). The charge sensitivity of this method is ~1.0 pC. This technique offers a simple and direct measurement to quantify the piezoelectric effect. However, the piezoelectric value might be affected by non-uniform stress distribution over the sample surface owing to the round shape of the loading contact. In addition, the thickness of the electrodes needs to be smaller than thin film thickness to avoid clamping effects which may cause deviation in the average value of the piezoelectric coefficient.

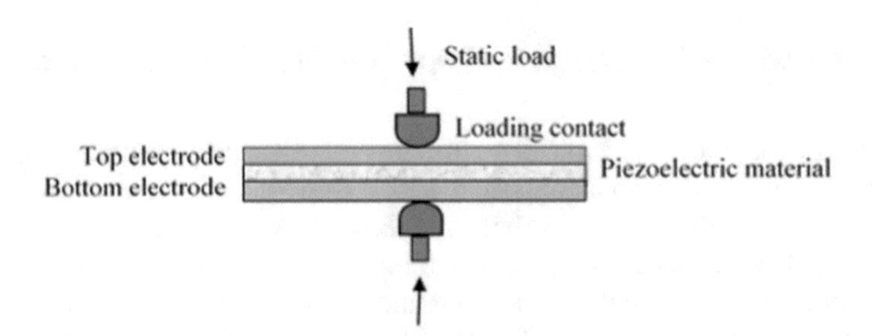

Fig. 2 Schematic drawing of quasi-static method

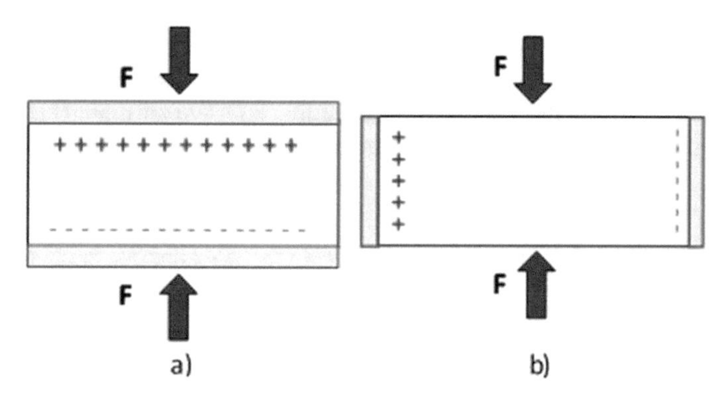

Fig. 3 Piezoelectric modes **a** d_{33} operating mode of energy harvester whereby the generated voltage is in the same direction as that of the applied force, *F* and **b** d_{31} whereby the generated voltage is perpendicular to the applied force, *F*

4.2 *Laser Interferometry*

The direct technique to quantify the piezoelectricity in thin-film bionanomaterial employs a polarized laser beam that is directed to the piezoelectric sample surface and reflected to the beam splitter (Fig. 4). The interference pattern is then produced due to a recombination of the reflected beam from a laser with the reflected beam from the reference mirror, which caused the sample to be vibrated at relatively low-frequency driven by AC signal. The interferometric change in the system is detected by a photodiode and monitored by a lock-in amplifier. The piezoelectric coefficient with high reliability data will be produced owing to the high resolution of interferometry in detecting the sample surface displacement ($<10^{-3}$ nm). The main source of inaccuracy using this method is the interference vibrations in the surroundings

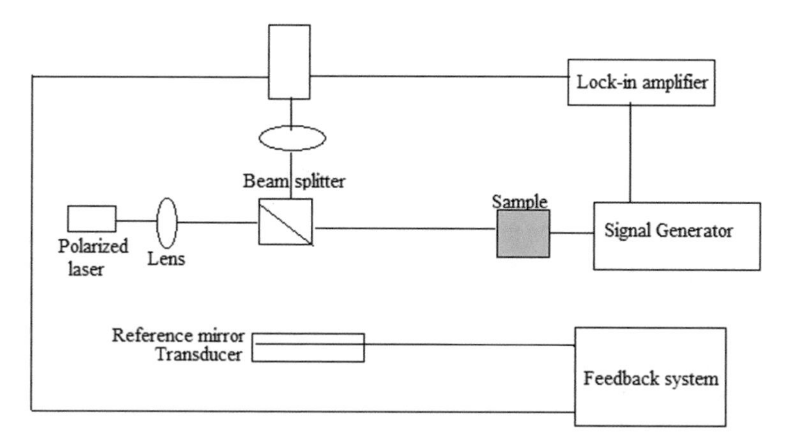

Fig. 4 Instrumental setup of laser interferometry

of the experimental workplace such as noise in the output interferometer signal. Lesser deviation in the dimensions of individual samples can have a big influence. The drawback of using this method is the clamping effect which directly disturbs the mode of vibration, subsequently reducing the piezoelectric coefficient.

4.3 *Frequency Measurement*

This indirect method is often used to quantify the piezoelectricity of thin film structure. This method depends on the frequency resonance response observed by the impedance analyzer. It involves the measurement of resonance and antiresonance frequency which correspond to the minimum and maximum impedance of a piezoelectric material [44]. As the thin-film structure's thickness is smaller than its width and length, the measurement of piezoelectric constant is gaining in transverse mode, d_{31}. The transverse piezoelectric coefficient employing frequency measurement can be calculated from:

$$d_{31} = \frac{1}{2\pi f_r} \sqrt{\frac{\varepsilon_{33}}{\rho\left[1 + \frac{8}{\pi^2}\left(\frac{f_r}{2(f_r - f_a)}\right)\right]}} \tag{6}$$

where f_r is the resonance frequency, f_a is the antiresonance frequency, ρ is the density of piezoelectric material, and ε_{33} is the permittivity of piezoelectric material, respectively. This method is complicated since it relies on the geometry of the materials and has relatively low resolution as compared to other methods. The frequency measurement setup is schematically presented in Fig. 5. The function of each instrument is summarized in Table 1.

4.4 *Piezoresponse Force Microscopy*

Piezoresponse force microscopy (PFM) adopts the piezoelectric converse effect where mechanical response is measured as an electric field is applied. It is known as the most sophisticated tool to measure the piezoelectric coefficient of nanomaterials. In PFM, the strain is induced due to the applied electric field. AC voltage is applied between the conductive tip and piezoelectric thin-film sample (Fig. 6). Subsequently, piezoelectric domains expand or contract due to the bias signal and generate in-phase response or out-phase response of piezoelectric domains. The amplitude of response is a quantitative measure for piezoelectric charge constant (d_{33}). To date, PFM has been acknowledged as a robust tool to quantify piezoresponse of nanostructures. It also can provide information on topography, domain size, and switching behavior of piezoelectric domains. A prime advantage of PFM is that it provides high data

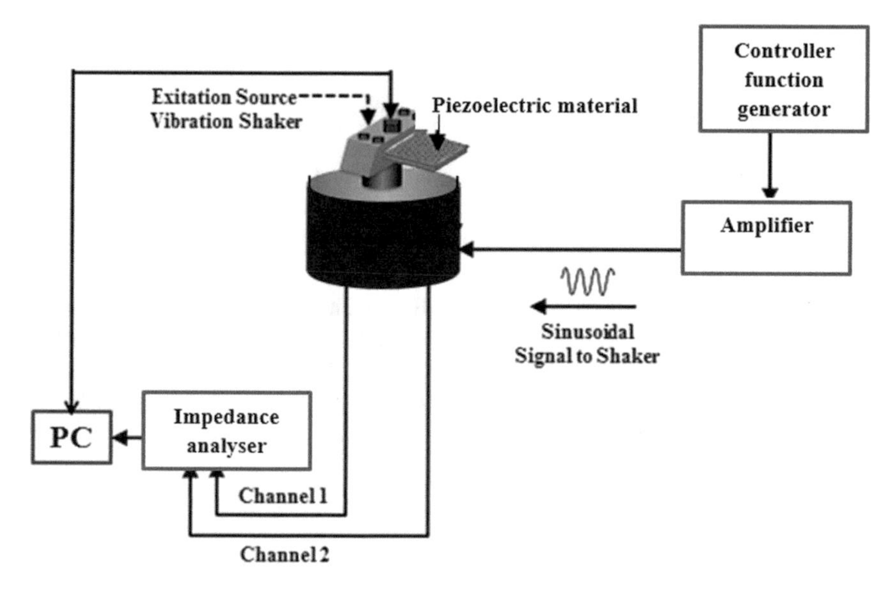

Fig. 5 Schematic illustration of frequency measurement setup

Table 1 Function of equipment in frequency measurement setup

Description	Function
Function generator and amplifier	To generate the signal to the shaker and to set frequency range
Vibration shaker	To produce vibration based on the signal from amplifier
Impedance analyzer	To observe the output frequency signal
Data acquisition software	To display the acceleration and voltage output

Fig. 6 Principal setup of PFM

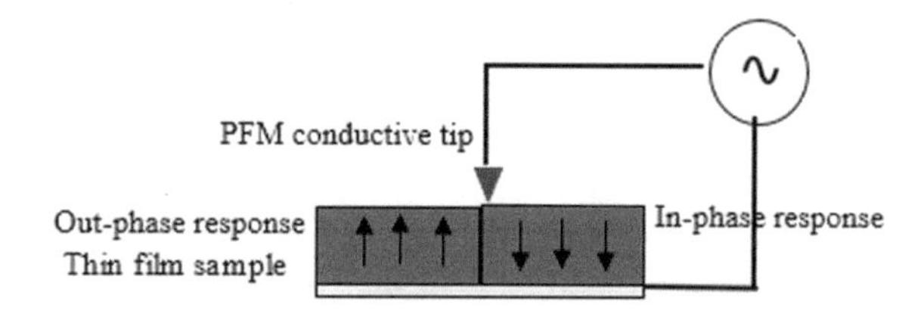

accuracy and reliability. Detailed information on the piezoelectric domain behavior at nanoscale can be obtained by simply applying the electric field to PFM conductive field.

4.5 Acoustic Wave Response

The acoustic wave technique is mainly developed for piezoelectric thin film deposited on a substrate as a transducer. Bulk acoustic wave (BAW) or surface acoustic wave (SAW) will be generated on the electrode controlled by AC electrical field. The disparity between BAW and SAW is the way that the wave pattern is generated. BAW moves from the top electrode along the longitudinal direction subsequently toward the finite depth of substrate. By contrast, SAW moves from one end of the electrode continuing along the film surface layer and finally being received by pattern electrode at the other end. In the electromechanical coupling coefficient, k is a vital parameter to quantify the piezoelectricity in thin film. It defines the change in velocity when the wave passes the conductive thin film [45]. From k, piezoelectric coefficient can be calculated using Eq. (7):

$$d_{33} = \sqrt{\varepsilon_{33} s_{33} k_{\mathrm{t}}^2} \tag{7}$$

where ε_{33} denotes dielectric permittivity of thin film and s_{33} is an elastic compliance, respectively. Electromechanical coupling factor is obtained using the derived equations according to Smith's equivalent circuit model [46].

4.6 X-Ray Diffraction Technique

It is a new indirect technique to measure piezoelectricity in piezoelectric materials [47]. The strains in a piezoelectric thin film induced by an electrical field can be measured using X-ray diffraction approach. d_{33} is estimated from the change in the out-of-plane lattice spacing of the piezoelectric thin film while d_{31} is measured from the change in the slope of the d spacing versus $\sin^2 \theta$ curves before and after applying an electric field over the films. Using this technique, the piezoelectric constant can be expressed as:

$$d_{33} = \Delta s_{33} / E_3 \tag{8}$$

where Δs_{33} is as strain tensor and E_{33} is an applied electric field. A principal advantage of these techniques is that they eliminate the use of elastic modulus in estimating the piezoelectric coefficient. Notably, the elastic modulus of thin films could be different from the bulk-type materials which in most cases is not known [48]. Since the converse piezoelectric coefficient considered the strains produced within the intrinsic structure of thin film rather than the displacement of thin films, this method also offers high sensitivity and high reliability of data due to the elimination of environment or vibration effects. Nevertheless, the geometry of top electrode would affect the piezoelectric coefficient of piezoelectric materials.

A summary of important tools that could be used in quantifying the piezoelectric effect in thin film in bionanomaterials is shown in Table 2. The operating principle and advantage or disadvantage of these tools are also included.

Table 2 Important tools in quantifying the piezoelectricity in bionanomaterial

Method and equipment	Principle operation	Properties measured	Strength/weakness	References
Quasi-static method (Berlincourt piezometer)	Directly apply the force to the piezoelectric materials and measuring the generated charge	Piezoelectric charge coefficient d_{33}/d_{31} (pC/N)	• Simple and direct measurement, sensitivity of piezometer requires the quality of electrode attached • Low resolution, clamping effect	[49–52]
Laser interferometry (Single/double beam interferometry)	The reflected beam is modulated by surface displacement whereby the electric field induced the displacement and inverse piezoelectric effect will be produced	Converse piezoelectric coefficient d_{33}/d_{31} (pm/V)	• High resolution and sensitivity, high reliability of data • Clamping effect	[53, 54]
Frequency measurement (Impedance analyzer)	Measurement of resonant and anti-resonant frequency which corresponds to the maximum or minimum impedance of piezoelectric material	Resonance and antiresonance frequency (f_a and f_r) Piezoelectric charge coefficient d_{33}/d_{31} (pC/N)	• Relatively low resolution, indirect measurement, geometry, and frequency-dependent, low data reliability	[55–58]
Piezoresponse force microscopy (PFM)	The surface displacement is modulated by an applied electric field from the tip of a scanning force microscope	Converse piezoelectric coefficient d_{33}/d_{31} (pm/V)	• Extremely high resolution and sensitivity, high reliability of data, can map the ferroelectric domain patterns with a lateral resolution of a few nanometers	[59–61]

(continued)

Table 2 (continued)

Method and equipment	Principle operation	Properties measured	Strength/weakness	References
Cantilever (Unimorph/bimorph cantilever structure)	Strain-induced charge/voltage	Transverse piezoelectric constant d_{31} (pm/V)	• Direct and simple, high sensitivity • Geometry dependent	[62, 63]
Acoustic wave response	Ultrasonic vibration-induced signal	Converse piezoelectric coefficient d_{33}/d_{31} (pm/V)	• Low reliability of date, geometry dependent, low resolution	[64–67]
X-ray diffraction	The strains in a piezoelectric thin film induced by an electrical field	Converse piezoelectric coefficient d_{33}/d_{31} (pm/V)	• Measure the piezoelectric coefficient without the requirement of elastic modulus, high sensitivity and high reliability of data • Clamping effect, the effect of geometry	[47]

5 Conclusion

In the last few years, piezoelectric principle in bionanomaterials has been practically demonstrated. Piezoelectric effect links the mechanical effects (stress and strain) and the electrical effects (voltage or charge). Interestingly, bionanomaterials such as cellulose, chitin, chitosan, amino acid, and peptide can possibly be utilized as flexible and wearable electronics due to substantial piezoelectric effects observed in these materials. A basic theory behind the piezoelectricity is discussed, primarily considering the existence of non-centrosymmetric structure of bionanomaterials. Further, the specified piezoelectric materials with outstanding piezoelectric properties are discussed, make them potentially preferable to power electronic devices and embedded system. The bionanomaterials discussed in this review are expected to have comparable properties to the market-dominant lead-based material, PZT, hence becoming a promising alternative for future electronic devices. The state of the art in quantifying piezoelectricity in bionanomaterial should not be neglected as it becomes a vital factor to drive the piezoelectric materials to the heights never envisaged before.

Acknowledgements The authors would like to acknowledge financial assistance from the Ministry of Education under the Fundamental Research Grant Scheme (FRGS) (FRGS/1/2019/TK02/UIAM/02/3).

References

1. M. Jacob, K.T. Varughese, S. Thomas, Dielectric characteristics of sisal-oil palm hybrid biofibre reinforced natural rubber biocomposites. J. Mater. Sci. **41**, 5538–5547 (2006). https://doi.org/10.1007/s10853-006-0298-y
2. T. Katsuura, S. Izumi, M. Yoshimoto, H. Kawaguchi, S. Yoshimoto, T. Sekitani, Wearable pulse wave velocity sensor using flexible piezoelectric film array, in *2017 IEEE Biomedical Circuits and Systems Conference (BioCAS)* (IEEE, 2017), pp. 1–4
3. J. Jacob, N. More, K. Kalia, G. Kapusetti, Piezoelectric smart biomaterials for bone and cartilage tissue engineering. Inflamm. Regen. **38**, 2–11 (2018). https://doi.org/10.1186/s41232-018-0059-8
4. M. Pohanka, Overview of piezoelectric biosensors, immunosensors and DNA sensors and their applications. Materials **11**, 448–461 (2018). https://doi.org/10.3390/ma11030448
5. M.T. Chorsi, E.J. Curry, H.T. Chorsi, R. Das, J. Baroody, P.K. Purohit, H. Ilies, T.D. Nguyen, Piezoelectric biomaterials for sensors and actuators. Adv. Mater. **31**, 237–274 (2019). https://doi.org/10.1002/adma.201802084
6. K. Kim, M. Ha, B. Choi, S.H. Joo, H.S. Kang, J.H. Park, B. Gu, C. Park, J. Kim, S.K. Kwak, Biodegradable, electro-active chitin nanofiber films for flexible piezoelectric transducers. Nano Energy **48**, 275–283 (2018)
7. A. Hänninen, E. Sarlin, I. Lyyra, T. Salpavaara, M. Kellomäki, S. Tuukkanen, Nanocellulose and chitosan-based films as low cost, green piezoelectric materials. Carbohydr. Polym. **202**, 418–424 (2018). https://doi.org/10.1016/j.carbpol.2018.09.001
8. F. Ali, W. Raza, X. Li, H. Gul, K.H. Kim, Piezoelectric energy harvesters for biomedical applications. Nano Energy **57**, 879–902 (2019). https://doi.org/10.1016/j.nanoen.2019.01.012
9. B.W. An, J.H. Shin, S.Y. Kim, J. Kim, S. Ji, J. Park, Y. Lee, J. Jang, Y.G. Park, E. Cho, S. Jo, Smart sensor systems for wearable electronic devices. Polymers **9**, 303–341 (2017). https://doi.org/10.3390/polym9080303
10. N.A. Hoque, P. Thakur, P. Biswas, M.M. Shaikh, S. Roy, B. Bagchi, S. Das, P.P. Ray, Biowaste crab shell-extracted chitin nanofiber-based superior piezoelectric nanogenerator. J. Mater. Chem. A **6**, 13848–13858 (2018). https://doi.org/10.1039/C8TA04074E
11. S.K. Ghosh, D. Mandal, High-performance bio-piezoelectric nanogenerator made with fish scale. Appl. Phys. Lett. **109**, 103701 (2016). https://doi.org/10.1063/1.4961623
12. S.K. Ghosh, D. Mandal, Efficient natural piezoelectric nanogenerator: electricity generation from fish swim bladder. Nano Energy **28**, 356–365 (2016). https://doi.org/10.1016/j.nanoen.2016.08.030
13. R. Dorey, R. Whatmore, Electroceramic thick film fabrication for MEMS. J. Electroceram. **12**, 19–32 (2004). https://doi.org/10.1023/B:JECR.0000033999.74149.a3
14. N.R. Harris, M. Hill, R. Torah, R. Townsend, S. Beeby, N.M. White, J. Ding, A multilayer thick-film PZT actuator for MEMs applications. Sens. Actuators A Phys. **132**, 311–316 (2006). https://doi.org/10.1016/j.sna.2006.06.006/
15. D. Kuščer, M. Skalar, J. Holc, M. Kosec, Processing and properties of $0.65Pb(Mg_{1/3}Nb_{2/3})O_3$–$0.35PbTiO_3$ thick films. J. Eur. Ceram. Soc. **29**, 105–113 (2009). https://doi.org/10.1016/j.jeurceramsoc.2008.06.010
16. S. Guerin, S.A.M. Tofail, D. Thompson, Organic piezoelectric materials: milestones and potential. NPG Asia Mater. **11**, 1–5 (2019). https://doi.org/10.1038/s41427-019-0110-5
17. M.A.M. Harttar, M.W.A. Rashid, U.A.A. Azlan, Physical and electrical properties enhancement of rare-earth doped-potassium sodium niobate (KNN): a review. Ceram.–Silikáty **59**, 158–163 (2015)
18. A.L. Kholkin, N.A. Pertsev, A.V. Goltsev, Piezoelectricity and crystal symmetry, in *Piezoelectric and Acoustic Materials for Transducer Applications*, ed. by A. Safari, E.K. Akdoğan (Springer, Boston, MA, 2008)

19. C.M. Weng, C.C. Tsai Hong, C.C. Lin, C.C. Chen, S.Y. Chu, J. Sheen, Z.Y. Chen, H.H. Su, Effects of non-stoichiometry on the microstructure, oxygen vacancies, and electrical properties of KNN-based thin films. ECS J. Solid State Sci. Technol. **5**, 49–56 (2016). https://doi.org/10.1149/2.0291609jss
20. M.M. Akmal, A.R.M. Warikh, U.A.A. Azlan, M.A. Azam, S. Ismail, Effect of amphoteric dopant on the dielectric and structural properties of yttrium doped potassium sodium niobate thin film. Mater. Lett. **170**, 10–14 (2016). https://doi.org/10.1016/j.matlet.2016.01.135
21. M.M. Akmal, A.R.M. Warikh, U.A.A. Azlan, N.A. Azmi, M.S. Salleh, M.S. Kasim, Optimizing the processing conditions of sodium potassium niobate thin films prepared by sol-gel spin coating technique. Ceram. Int. **44**, 317–325 (2018). https://doi.org/10.1016/j.ceramint.2017.09.175
22. V.A.D. De Almeida, F.G. Baptista, An experimental assessment of PZT patches for impedance-based SHM applications, in *Proceedings of the International Conference on Sensing Technology, ICST* (2014), pp. 495–499
23. N.M. Hagh, K. Kerman, B. Jadidian, A. Safari, Dielectric and piezoelectric properties of Cu^{2+}-doped alkali niobates. J. Eur. Ceram. Soc. **29**, 2325–2332 (2009). https://doi.org/10.1016/j.jeurceramsoc.2009.01.003
24. L. Csoka, I.C. Hoeger, O.J. Rojas, I. Peszlen, J.J. Pawlak, P.N. Peralta, Piezoelectric effect of cellulose nanocrystals thin films. ACS Macro Lett. **1**, 867–870 (2012). https://doi.org/10.1021/mz300234a
25. F.B. Ahmad, Z. Zhang, W.O.S. Doherty, I.M. O'Hara, The prospect of microbial oil production and applications from oil palm biomass. Biochem. Eng. J. **143**, 9–23 (2018). https://doi.org/10.1016/j.bej.2018.12.003
26. H. Du, W. Liu, M. Zhang, C. Si, X. Zhang, B. Li, Cellulose nanocrystals and cellulose nanofibrils based hydrogels for biomedical applications. Carbohydr. Polym. **209**, 130–144 (2019). https://doi.org/10.1016/j.carbpol.2019.01.020
27. F. Rol, M.N. Belgacem, A. Gandini, J. Bras, Recent advances in surface-modified cellulose nanofibrils. Prog. Polym. Sci. **88**, 241–264 (2019). https://doi.org/10.1016/j.progpolymsci.2018.09.002
28. S. Rajala, M. Vuoriluoto, O.J. Rojas, S. Franssila, S. Tuukkanen, Piezoelectric sensitivity measurements of cellulose nanofibril sensors, in *IMEKO XXI World Congress, Proceedings*, Prague, Czech Republic, 30 Aug–4 Sept 2015 (2015)
29. S.K. Karan, S. Maiti, S. Paria, A. Maitra, S.K. Si, J.K. Kim, B.B. Khatua, A new insight towards eggshell membrane as high energy conversion efficient bio-piezoelectric energy harvester. Mater. Today Energy **9**, 114–125 (2018). https://doi.org/10.1016/j.mtener.2018.05.006
30. S.K. Ghosh, D. Mandal, Sustainable energy generation from piezoelectric biomaterial for noninvasive physiological signal monitoring. ACS Sustain. Chem. Eng. **5**, 8836–8843 (2017). https://doi.org/10.1021/acssuschemeng.7b01617
31. A. Hänninen, S. Rajala, T. Salpavaara, M. Kellomäki, S. Tuukkanen, Piezoelectric sensitivity of a layered film of chitosan and cellulose nanocrystals. Procedia Eng. **168**, 1176–1179 (2016). https://doi.org/10.1016/j.proeng.2016.11.397
32. S. Rajala, T. Siponkoski, E. Sarlin, M. Mettänen, M. Vuoriluoto, A. Pammo, J. Juuti, O.J. Rojas, S. Franssila, S. Tuukkane, Cellulose nanofibril film as a piezoelectric sensor material. ACS Appl. Mater. Interfaces **8**, 15607–15614 (2016). https://doi.org/10.1021/acsami.6b03597
33. R. Mangayil, S. Rajala, A. Pammo, E. Sarlin, J. Luo, V. Santala, M. Karp, S. Tuukkanen, Engineering and characterization of bacterial nanocellulose films as low cost and flexible sensor material. ACS Appl. Mater. Interfaces **9**, 19048–19056 (2017). https://doi.org/10.1021/acsami.7b04927
34. E. Praveen, S. Murugan, K. Jayakumar, Investigations on the existence of piezoelectric property of a bio-polymer—chitosan and its application in vibration sensors. RSC Adv. **7**, 35490–35495 (2017). https://doi.org/10.1039/c7ra04752e
35. F.B. Ahmad, M.H. Maziati Akmal, A. Amran, M.H. Hasni, Characterization of chitosan from extracted fungal biomass for piezoelectric application. IOP Conf. Ser. Mater. Sci. Eng. **778**, 012034 (2020)

36. J. Jin, D. Lee, H.G. Im, Y.C. Han, E.G. Jeong, M. Rolandi, K.C. Choi, B.S. Bae, Chitin nanofiber transparent paper for flexible green electronics. Adv. Mater. **28**, 5169–5175 (2016). https://doi.org/10.1002/adma.201600336
37. H. El Knidri, R. Belaabed, A. Addaou, A. Laajeb, A. Lahsini, Extraction, chemical modification and characterization of chitin and chitosan. Int. J. Biol. Macromol. **120**, 1181–1189 (2018). https://doi.org/10.1016/j.ijbiomac.2018.08.139
38. R.M. Street, T. Huseynova, X. Xu, P. Chandrasekaran, L. Han, W.Y. Shih, W.H. Shih, C.L. Schauer, Variable piezoelectricity of electrospun chitin. Carbohydr. Polym. **195**, 218–224 (2018). https://doi.org/10.1016/j.carbpol.2018.04.086
39. S.K. Ghosh, D. Mandal, Bio-assembled, piezoelectric prawn shell made self-powered wearable sensor for non-invasive physiological signal monitoring. Appl. Phys. Lett. **110**, 123701 (2017). https://doi.org/10.1063/1.4979081
40. W.M. Nawawi, M. Jones, R.J. Murphy, K.Y. Lee, E. Kontturi, A. Bismarck, Nanomaterials derived from fungal sources—is it the new hype? Biomacromolecules **21**, 30–55 (2019). https://doi.org/10.1021/acs.biomac.9b01141
41. C.C. Silva, C.G.A. Lima, A.G. Pinheiro, J.C. Góes, S.D. Figueiró, A.S.B. Sombra, On the piezoelectricity of collagen–chitosan films. Phys. Chem. Chem. Phys. **3**, 4154–4157 (2001). https://doi.org/10.1039/B100189M
42. A.H. Rajabi, M. Jaffe, T.L. Arinzeh, Piezoelectric materials for tissue regeneration: a review. Acta Biomater. **24**, 12–23 (2015). https://doi.org/10.1016/j.actbio.2015.07.010
43. V.V. Lemanov, Ferroelectric and piezoelectric properties of protein amino acids and their compounds. Phys. Solid State **54**, 1841–1842 (2012). https://doi.org/10.1134/S1063783412090168
44. H. Yuan, T. Lei, Y. Qin, J.H. He, R. Yang, Design and application of piezoelectric biomaterials. J. Phys. D Appl. Phys. **52**, 194002 (2019). https://doi.org/10.1088/1361-6463/ab0532
45. K. Jenkins, S. Kelly, V. Nguyen, Y. Wu, R. Yang, Piezoelectric diphenylalanine peptide for greatly improved flexible nanogenerators. Nano Energy **51**, 317–323 (2018). https://doi.org/10.1016/j.nanoen.2018.06.061
46. V. Nguyen, R. Zhu, K. Jenkins, R. Yang, Self-assembly of diphenylalanine peptide with controlled polarization for power generation. Nat. Commun. **7**, 1–6 (2016). https://doi.org/10.1038/ncomms13566
47. J. Cheeke, Y. Zhang, Z. Wang, M. Lukacs, M. Sayer, Characterization for piezoelectric films using composite resonators, in *1998 IEEE Ultrasonics Symposium. Proceedings (Cat. No. 98CH36102)* (IEEE, 1998), pp. 1125–1128
48. A.F. Malik, V. Jeoti, M. Fawzy, A. Iqbal, Z. Aslam, M.S. Pandian, E. Marigo, Estimation of SAW velocity and coupling coefficient in multilayered piezo-substrates AlN/SiO_2/Si, in *2016 6th International Conference on Intelligent and Advanced Systems (ICIAS)* (IEEE, 2016), pp. 1–5
49. V.T. Rathod, A review of electric impedance matching techniques for piezoelectric sensors, actuators and transducers. Electronics **8**, 169–174 (2019). https://doi.org/10.3390/electronics8020169
50. Y.H. Yu, M.O. Lai, L. Lu, Measurement of thin film piezoelectric constants using X-ray diffraction technique. Phys. Scr. **129**, 353–357 (2007). https://doi.org/10.1088/0031-8949/2007/t129/078
51. S. Bühlmann, B. Dwir, J. Baborowski, P. Muralt, Size-effect in mesoscopic epitaxial ferroelectric structures: increase of piezoelectric response with decreasing feature-size. Integr. Ferroelectr. **50**, 261–267 (2002). https://doi.org/10.1080/743817662
52. M.A.M. Hatta, M.W.A. Rashid, U.A.A. Azlan, K.S. Leong, N.A. Azmi, Finite element method simulation of MEMS piezoelectric energy harvester using lead-free material, in *2016 International Conference on Computer and Communication Engineering (ICCCE)* (IEEE, 2016), pp. 511–515
53. J. Fialka, P. Beneš, Comparison of methods of piezoelectric coefficient measurement, in *2012 IEEE International Instrumentation and Measurement Technology Conference Proceedings*, 13–16 May 2012, pp. 37–42

54. M. Stewart, M.G. Cain, P. Weaver, Electrical measurement of ferroelectric properties, in *Characterisation of Ferroelectric Bulk Materials and Thin Films*, ed. by M. Cain. Springer Series in Measurement Science and Technology, vol. 2 (Springer, Dordrecht, 2014)
55. M.G. Cain, *Characterisation of Ferroelectric Bulk Materials and Thin Films*, vol. 2 (Springer, 2014)
56. Z. Zhao, H.L.W. Chan, C.L. Choy, Determination of the piezoelectric coefficient d_{33} at high frequency by laser interferometry. Ferroelectrics **195**, 35–38 (1997). https://doi.org/10.1080/00150199708260482
57. A. Gruverman, M. Alexe, D. Meier, Piezoresponse force microscopy and nanoferroic phenomena. Nat. Commun. **10**(1), 1–9 (2019). https://doi.org/10.1038/s41467-019-09650-8
58. H. Zhou, R.H. Han, M.H. Xu, H. Guo, Study of a piezoelectric accelerometer based on d_{33} mode, in *2016 Symposium on Piezoelectricity, Acoustic Waves, and Device Applications* (IEEE, 2016), pp. 216–219
59. R. Mohamed, M.R. Sarker, A. Mohamed, An optimization of rectangular shape piezoelectric energy harvesting cantilever beam for micro devices. Int. J. Appl. Electromagn. Mech. **50**, 537–548 (2016). https://doi.org/10.3233/jae-150129
60. I. Kanno, H. Kotera, K. Wasa, Measurement of transverse piezoelectric properties of PZT thin films. Sens. Actuator A Phys. **10**, 68–74 (2009)
61. W. Sriratana, R. Murayama, L. Tanachaikhan, Synthesis and analysis of PZT using impedance method of reactance estimation. Adv. Mater. Phys. Chem. **3**, 62–70 (2013)
62. L. Li, Y. Yang, Z. Liu, S. Jesse, S.V. Kalinin, R.K. Vasudevan, Correlation between piezoresponse nonlinearity and hysteresis in ferroelectric crystals at the nanoscale. Appl. Phys. Lett. **108**, 172905 (2016). https://doi.org/10.1063/1.4947533
63. H. Shin, J. Song, Piezoelectric coefficient measurement of AlN thin films at the nanometer scale by using piezoresponse force microscopy. J. Korean Phys. Soc. **56**, 580–585 (2010)
64. D.A. Bonnell, S.V. Kalinin, A. Kholkin, A. Gruverman, Piezoresponse force microscopy: a window into electromechanical behavior at the nanoscale. MRS Bull. **34**, 648–657 (2009). https://doi.org/10.1557/mrs2009.176
65. H. Wang, X. Shan, T. Xie, Performance optimization for cantilevered piezoelectric energy harvester with a resistive circuit, in *2012 IEEE International Conference on Mechatronics and Automation* (IEEE, 2012), pp. 2175–2180
66. K. Wasa, I. Kanno, H. Kotera, Fundamentals of thin film piezoelectric materials and processing design for a better energy harvesting MEMS. Power MEMS **61**, 61–66 (2009)
67. L. Pamwani, A. Habib, F. Melandsø, B. Ahluwalia, A. Shelke, Single-input and multiple-output surface acoustic wave sensing for damage quantification in piezoelectric sensors. Sensors **18**, 1–19 (2017). https://doi.org/10.3390/s18072017

Experimental Methods for the Phytochemical Production of Nanoparticles

Fatemeh Soroodi, Parveen Jamal, and Ibrahim Ali Noorbatcha

Abstract In view of the numerous applications of the nanoparticles, the synthesis of nanoparticles continues to attract much attention. There are several methods available to synthesize nanoparticles. In general, these synthesis methods can be classified under physical, chemical, and biochemical methods and each of these methods has its own advantages and disadvantages. Green synthesis using plants provides an alternative approach for the synthesis of nanoparticles. Phytochemical method of nanoparticles synthesis offers advantages over other methods as it is simple, cost-effective, relatively reproducible and often results in more stable materials. The existing biomolecules and phytochemicals in the plant extracts are considered the reducing and capping factors in the formation of stable nanoparticles. In this paper, we review the use of plants and the effect of process parameters on the synthesis of nanoparticles. In addition, we have also described a method developed by us to synthesize selenium (Se) nanoparticles using cocoa pod husk extract.

Keywords Nanoparticles · Plants · Phytochemical synthesis · Selenium nanoparticle · Characterization

1 Introduction

Recently, nanotechnology has emerged as an elementary division of modern science and is receiving global attention due to its extensive applications. Nanomaterials controlled to nanocrystalline size (less than 100 nm) show different characteristics compared to their bulk equivalents. These unique properties of nanoparticles (NPs) depend on the atomic structure, size confinement, composition, microstructure, defects, and interfaces, all of which can be attempted by the method of synthesis [1]. Consequently, the techniques for NPs formation have turned out to be one of the most important parts in nanotechnology. Literature suggested that various NPs have been synthesized by different methods. In general, these methods include top-down

F. Soroodi · P. Jamal · I. A. Noorbatcha (✉)
Department of Biotechnology Engineering, Kulliyyah of Engineering, International Islamic University Malaysia, 50728 Kuala Lumpur, Malaysia
e-mail: noorbatcha@gmail.com

A. T. Jameel and A. Z. Yaser (eds.), *Advances in Nanotechnology and Its Applications*,
https://doi.org/10.1007/978-981-15-4742-3_5

and bottom-up approaches [2] which practically contain physical, chemical, and biological techniques. Various physical and chemical methods have been employed to synthesize nanomaterials with different properties. The advantage of the physical methods over the chemical ones is the lack of chemical reagent at the time of NPs formation while the disadvantage is the need for providing high amount of energy for the commercial scale production [3].

In the last two decades, biosynthesis of nanoparticles has been highly investigated [4]. The biological or green synthesis route is very spontaneous, economic, environmentally friendly, and non-toxic. Moreover, this method can provide NPs of a better-defined size and morphology than some of the physicochemical route of synthesis [5]. Microorganisms including bacteria, fungi, yeast, actinomycetes, and algae by secreting proteins and enzymes have been highly known to hold great potential as effective and eco-friendly tools for biosynthesis of NPs, avoiding harsh chemical and high-energy demand [6–8]. Despite all the benefits, the contaminating nature of microorganisms as well as high cost for their maintenance need to be considered which are the drawbacks in comparison with the use of plant or plant extract for NPs synthesis. In phytochemical production, an abundant number of plants are being exploited as an intermediary to reduce various metal salts into nanoparticles as a nonpathogenic and one-step protocol.

2 Phytochemical Synthesis of Nanoparticles

Simplicity and the capability of mass production have led plant-mediated NPs synthesis to become a new trend for the production of NPs. Plant-derived NPs elevated by plant extracts commonly consists of three different phases [9], namely induction, growth, and termination. The first step is rapid metal ion reduction and nucleation which result in unstable and reactive particles. These small crystals grow and aggregate spontaneously until the shape and size of particles reach the desired energy level. This is followed by the termination phase when capping agents stabilize the NPs [10].

The fundamental mechanism regarded in the production of NPs assisted by the plants is as a result of the presence of phytochemicals. The major phytochemicals responsible for the spontaneous reduction of ions are flavonoids, alkaloids, terpenoids, carboxylic acids, aldehydes, amides, ketones, and quinones. It was also found that chloroplasts containing high amount of glucose and fructose as reducing sugars play a significant role in the reduction of metal ions to nanosize [11]. Apart from mediating the synthesis, these biomolecules also stabilize the NPs formed with desired size and shape as capping agents [12]. Zheng et al. investigated that *Cacumen platycladi* extract derived Pt NPs were biologically synthesized by the reducing sugars and flavonoid rather than proteins [13]. Whereas, Ag NPs formed by *Aloe vera* leaf extract showed capping property due to the adsorbed carboxyl of the present proteins to the metal NPs and securing them from aggregation [14].

Many studies have reported that a large number of NPs formation were successfully assisted by the extract prepared from different parts of plants including leaf, flower, seed, fruit, bark, root, rhizome, tuber, juice, latex, gall, waste such as peel, pulp, shell, and even the whole plant. Table 1 presents some examples of phytochemical production of NPs.

Generally, leaves are more favorable to researchers. It is because of the presence of chlorophylls and a big quantity of different phytochemicals in the extract. For

Table 1 Use of different parts of plants in phytochemical synthesis of NPs

Part of plant	Plant name	NPs	Size (nm)	Shape	References
Leaf	*Solanum nigrum*	Au	50	Spherical	[15]
	Eucalyptus spp.	Ag	4–60	Spherical	[16]
	Gloriosa superba L.	CeO	5	Spherical	[17]
	Agathosma betulina	CdO	8	Quasi-spherical	[18]
	Aloe vera	CuO	20	Spherical	[19]
	Tea	Fe_2O_3	4–5	Spherical	[20]
	Acalypha indica	RuO_2	6–25	Spherical	[21]
	Pongamia pinnata	ZnO	100	Spherical	[22]
Fruit	*Tanacetum vulgare*	Ag, Au	10–40	Different shapes	[23]
Flower	*Gnidia glauca*	Au	10	Spherical	[24]
Petals	*Rosa hybrida*	Au	10	Different shapes	[25]
Latex	*Jatropha curcas*	Ag	20–30	–	[26]
Seed	*Abelmoschus esculentus*	Au	45–75	Spherical	[27]
Rhizome	*Zingiber officinale*	Ag	5–15	–	[28]
Juice	*Actinidia deliciosa*	Ag	5–25	–	[29]
Bran	*Sorghum spp*	Ag Fe	10 50	– Amorphous	[30]
Bark	*Cinnamon zeylanicum*	Ag	31–40	Spherical	[31]
Gall	*Pistacia integerrima*	Au	20–200	–	[32]
Tuber	*Potato*	ZnO	1.2	Hexagonal	[33]
Plant	*Parthenium hysterophorus*	ZnO	27.84	Spherical and hexagonal	[34]
Aerial parts	*Stachys lavandulifolia* Vahl	Au	56.3	From spherical to triangular	[35]
Waste	*Vitis vinifera* L.	Au	20–25	Different shapes	[36]
Pulp	*Beta vulgaris*	Au	20–200	Different shapes	[37]
Nutshell	*Anacardium othonianum* (cashew)	Ag	–	Spherical	[38]
Peel	*Citrus limone*	Ag	17.3 and 61.2	Different shapes	[39]

instance, spherical Ag NPs with the size range between 4 and 60 nm were immediately formed by the addition of 50 mL of eucalyptus leave extract into 25 mL of 0.1 M $AgNO_3$ salt solution at 30 °C and 150 rpm of stirring [16]. Fruits, flowers or petals extract as a reducing agent can also be applied as fresh or dried in the production of NPs [23, 25].

Latex is capable of being directly used in NPs formation which makes it easy as no extract preparation is required [26]. Utilizing juice has also reported as a stable and facile approach for green synthesis of NPs [29].

Rhizomes and roots have also been used in the green synthesis of NPs. In one study, 1 mL of ginger (*Z. officinale*) extract was used to reduce 25 mL of 1 mM $HAuCl_4$ into Au NPs [28]. The extract was just prepared by washing, cutting and boiling 20 g of the fresh rhizome with 250 mL deionized water and finally filtering it. Some other parts of the plants such as seed, bark, tuber, barn, and even some annual plants are usually used in powder form [31].

Every year, a big quantity of by-product wastes such as peel are generated which causes environmental problems. Most of these wastes are a good source of the phyto-chemicals and can mediate the biosynthesis of NPs. In one survey, only 3 mL of lemon peel extract was added to 40 mL of 1 mM AgNO3 solution for Ag NPs formation at room temperature [39].

3 Factors Affecting Green Synthesis of Nanoparticles

The effect of environmental parameters on biosynthesis of NPs has been widely highlighted in the literature. These factors including concentration of plant extract, concentration of salt solution, temperature, pH, etc., influence the size distribution and morphology as well as the concentration of the produced NPs.

3.1 Temperature

One of the most attracting perspectives of the green synthesis of NPs refers to the fact that the process is carried out at room temperature. However, the increased tempera-ture affects the reaction rate resulting in a faster completion of the process [40–42]. Temperature modification can also change the size and concentration of differently shaped NPs. Au NPs prepared by the reduction of $AuCl_4^-$ ions by lemongrass extract at room temperature, 40, 50, 60, 70, and 80 °C showed that the average particle size decreases with increasing temperature [43]. Moreover, with increasing temper-ature, the percentage of triangles decreased while the number of spherical Au NPs increased.

3.2 Extract Concentration

Extract concentration is also considered to control size distribution and reaction rate of metal salts into NPs. The findings of a study revealed that high concentration of *Magnolia kobus* leaf caused an increase in the size of the produced Ag NPs due to the high amount of reducing agent, leading to the aggregation of NPs [41]. In another experiment, there was a shift from 35 to about 100% in the conversion rate of metal ions into NPs after changing *Aloe vera* extract concentration from 5 to 25% [44]. At concentrations higher than optimum, there is no change in the rate of conversion. However, a change in NPs properties was observed by Ghosh et al. [24].

3.3 Salt Solution Concentration

The concentration of the metal ion can influence both the concentration and the size of the formed NPs. The effect of $AgNO_3$ concentration on conversion and particle size with 5% Magnolia leaf broth was investigated by Song and Kim [41]. The result proved that lower concentrations (0.1 and 1 mM) $AgNO_3$ required less than 11 min for more than 90% conversion, whereas the reaction with 2 mM $AgNO_3$ completed within 90 min. Additionally, the Ag NPs average size was reported to increase with lower salt concentration. It is also expected to have maximum concentration of NPs in the optimum condition of the salt solution concentration. On the other hand, beyond this concentration, the formation of NPs is inhibited [42].

3.4 pH

The pH of the reaction mixture can influence the concentration of NPs under formation to a great extent. The biogenic synthesis of highly stable Au NPs using the aqueous extract of *Citrus lemon* treated various pH (2–10) indicated that the optimum yield was resulted at pH 2.5 [45]. Besides, the stability of the Au NPs was found to be very high when synthesized using acidic environment (pH 2.5), whereas at higher pH (pH 6–10) the increase in the size of the nanoparticles was observed indicated by the color changed from wine red to purple.

4 Methodology

The general protocol for the preparation of plant extract consists of collecting that specific plant or plant material and properly washing it to remove any kind of dust and debris. The way of plant extract preparation can differ in the use of fresh or dry parts of the plant. Then, the clean sources are extracted with proper amount of deionized distilled water. This step may or may not involve heating. Finally, the mixture is followed by filtering thoroughly until all undesired particles are removed from the extract. The prepared extract is stored at 4 °C and used during the week.

Cocoa pod husk (CPH) is the main byproduct generated by an economically important crop called *Theobroma cacao* L. Every year, substantial quantity of CPH is left to decompose in the plantation area which causes environmental problems. Studies showed that this waste contains some phytochemicals, enzyme, and polysaccharide [46–48].

The goal of this experiment is phytochemical production of selenium nanoparticles (Se NPs) from cocoa pod husk (CPH). Therefore, various steps in the procedure of CPH extract preparation and synthesis of Se NPs will be discussed.

4.1 Materials and Methods

The required materials, apparatuses, glassware, and consumables for the green production of Se NPs include cocoa pod husk (CPH), 1 mM sodium selenite (Na_2SeO_3) solution, deionized water, balance, blender, centrifuge machine and 50 mL conical centrifuge tubes, heater, magnetic stirrer, gradual cylinder, 100-mL-Erlenmeyer flask, aluminum foil, micropipette, UV-vis spectrophotometer and quartz cuvette, FESEM, and FTIR.

4.2 CPH Extract Preparation

In this study, 50% (w/v) CPH extract was prepared by thoroughly washing CPH with tap water to remove any dust and impurities followed by rinsing CPH samples with deionized water. The amount of 20 g of CPH was added to 40-mL deionized water and blended the mixture for 3 min. The mixture was centrifuged at 10,000 rpm for 30 min, and the clear supernatant as liquid CPH extract was collected. It is suggested to use the fresh extract as it might lose its bio-reduction ability.

4.3 Phytochemical Synthesis of Se NPs

The amount of 15 mL of 50% (w/v) CPH extract was added into a 100 mL Erlenmeyer flask containing 80 mL 1 mM Na_2SeO_3 and covered by aluminum foil. The reaction mixture was placed on heater (60 °C) and mixed vigorously using magnetic stirrer for 10 min. The temperature was then reduced to 30 °C for 48 h.

Control solutions are necessarily required in order to notice the color change of the reaction mixture, i.e., the formation of Se NPs. Therefore, it is suggested to provide two different control solutions by the addition of 15 mL of deionized water into a 100–mL-Erlenmeyer flask containing 80 mL 1 mM Na_2SeO_3 for non-extract containing control and 15 mL CPH into 80 mL deionized water for non-salt solution control.

4.4 Observation

The red color in the sample is the preliminary evidence displaying the formation of Se NPs. Towards the end of the incubation time, gradual explicit color change from yellowish-brown to brick red in the experimental reaction mixture was observed. On the other hand, the control remained without any changes in the color.

4.5 Washing and Purification

Characterization of the synthesized NPs requires purified samples. In this study, the mixture of plant extract and synthesized NPs was centrifuged at 14,500 rpm (24,482.44 $\times$g) for 30 min. The precipitant was collected and was further washed with deionized water to get rid of the free proteins that are not capping the NPs. During the washing interval, samples were vortexed and sonicated for 15 min to disperse NPs in the solution. A dried powder of the NPs was obtained by freeze-drying technique.

4.6 UV-Vis Spectrophotometry

The measurement was carried out in the range of wavelengths from 200 to 800 nm and 1 cm for path length (Thermo Scientific, Fiberlite F15-8x50cy). Two mL of each control solutions were used as blank followed by measuring 2 mL of the experimental sample.

4.7 Field Emission Scanning Electron Microscope (FESEM)

FESEM (JSM-6700F, Japan) was used for the observation of the shape and approximate size of the synthesized Se NPs. The freeze-dried form of Se NPs was placed on a carbon tape and coated for the size and shape analysis. The sample was viewed at an operating accelerating voltage at 10 kV and various resolutions.

4.8 Fourier Transform Infrared Spectroscopy (FTIR)

The measurement was carried out between the range 500 and 4000 nm using FTIR (Thermo Scientific, FTIR Is50) at the resolution of 4 cm^{-1}. A thin film of freeze-dried Se NPs sample was placed between KBr and NaCl plates. The emitted infrared energy is transmitted upon entering the sample compartment and the unique frequency of energy characteristic of the sample is absorbed.

5 Results and Discussion

The addition of CPH extract to 1 mM liquid sodium selenite led to visual red color change and resulted in the reduction of selenite ions to Se NPs during the incubation period. According to Zhang et al., the shape of the Se NPs can be recognized by their color since red Se colloids can imply the amorphous NPs while trigonal Se NPs are black [49].

Figure 1 shows the recorded absorption spectra of suspended Se NPs in the mixture that exhibited a sharp peak at 221 nm which is in the characteristic range of Se NPs [50]. The presence of the large blue shift in the spectrum corresponds to the crystallizability of Se and the small size of the particles. Based on Mie's theory, spherical NPs reveals a single SPR band while anisotropic particles represent two or more peaks depending on the shape [51].

According to FESEM image (Fig. 2), the synthesized Se NPs demonstrate the particle size of less than 100 nm. The average size of Se NPs was found to be in the range of 25–45 nm. The variable size distribution of the particle suggests that CPH extract could form poly-disperse NPs [52]. Other factors such as concentration, pH, the nature of the present biomolecules, and temperature affect both shape and size of Se NPs [53].

FTIR spectrum study was performed to investigate the possible bio-reducing functional groups or capping agents present in CPH extract. The results of FTIR analysis of the CPH extract (i.e., the control solution) and the synthesized Se NPs are presented in Fig. 3a, b, respectively (Fig. 3). The FTIR spectrum shows that the broad absorption

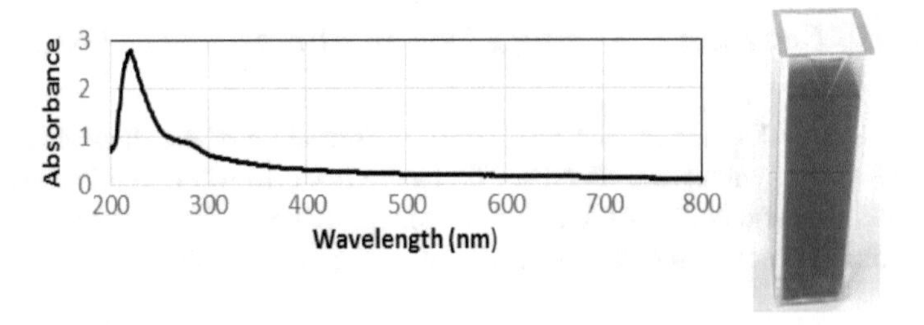

Fig. 1 CHP mediated Se NPs in aqueous solution after washing and the UV-vis absorption spectrum

Fig. 2 FESEM image of CPH mediated Se NPs

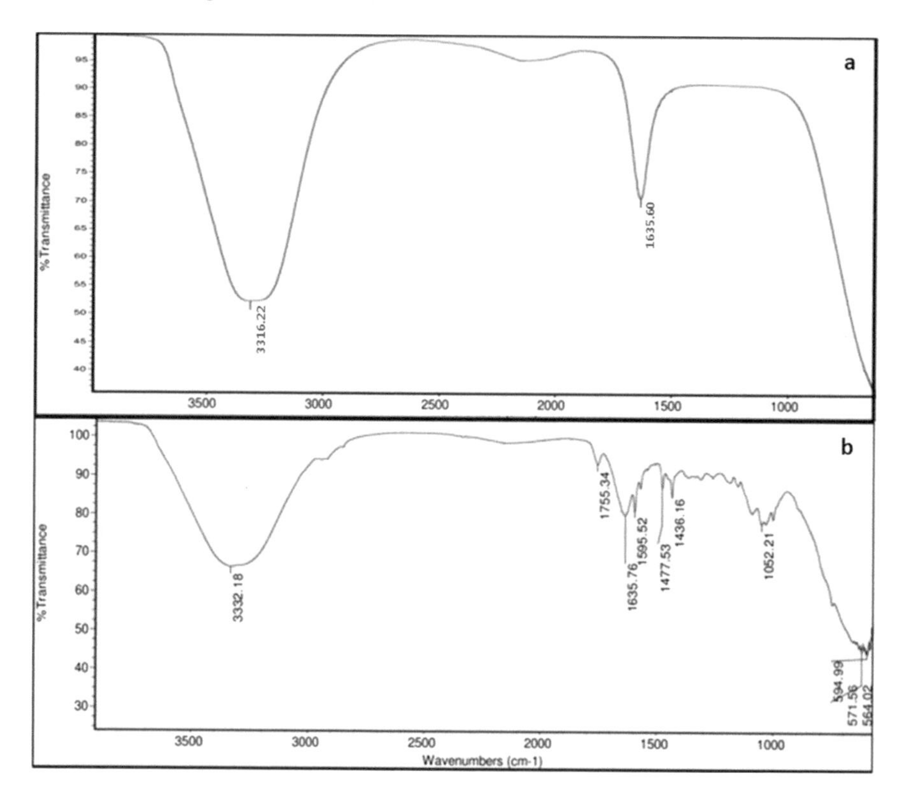

Fig. 3 FTIR spectra of CPH extract control solution (**a**) and CPH mediated Se NPs (**b**)

peak at 3316 cm^{-1} in CPH extract shifted to 3332 cm^{-1} in Se NPs with the increase in the intensity and can be assigned for stretching of N–H (amide) and O–H (of alcohols or phenols) groups. The transmittance peak positioned at 1635 cm^{-1} corresponding to N–H and C=O (amide) group in Se NPs was about 10 cm^{-1} higher than CPH extract. These transformations suggest that there is an interaction between Se NPs and hydroxyl and amino groups which might reduce $(SeO_3)^{-2}$ ions into Se^0. In addition to the peak 1635 cm^{-1}, new peaks were identified at 1755, 1595, and 1436 cm^{-1} with increase in the intensity which may be due to the modification of any previously existing functional groups on the surface of NPs [54].

The N–H stretch and bend indicate that protein might play a significant role in CPH-mediated Se NPs synthesis. The protein content of CPH was reported to be in the range of 6.8–10% w/w [46] which should be taken into consideration in process involving CPH. Since CPH contains carbohydrates such as pectin [47], the presence of O–H bond absorption frequency, a hemiacetal reducing end, and other functionalities in the polysaccharides can play important roles in both the reduction and the stabilization of metallic nanoparticles [55].

6 Conclusion

In conclusion, our observations indicate that CPH extract has reducing property to mediate the synthesis of selenium ions into Se NPs. It is suggested that the existing phytochemicals and enzymes in the extract are the reagents responsible for the biosynthesis of Se NPs.

7 Suggestion

It is important to verify the chemical and physical properties of NPs that are obtained from the experiment. In this respect, many characterization methods have been developed in order to help researchers in their studies. Zetasizer, scanning electron microscopy (SEM), energy-dispersive X-ray spectroscopy (EDX), and X-ray diffraction (XRD) are some of the common characterization techniques applied in NPs synthesis which requires relative sample preparation.

References

1. Y. Huang, J.J. McCormick, J. Economy, Adhesion of an aromatic thermosetting copolyester with copper foils. Polym. Adv. Technol. **16**(1), 1–5 (2005)
2. F. Bensebaa, Chapter 3—Dry production methods, in *Interface Science and Technology*, ed. by B. Farid, vol. 19 (2013), pp. 147–184

3. T. Panigrahi, *Synthesis and characterization of silver nanoparticles using leaf extract of Azadirachta indica*, Doctoral dissertation, National Institute of Technology Rourkela, 2013
4. T.M. Abdelghany, A.M.H. Al-Rajhi, M.A. Al Abboud, M.M. Alawlaqi, A. Ganash Magdah, E.A.M. Helmy, A.S. Mabrouk, Recent advances in green synthesis of silver nanoparticles and their applications: about future directions. A review. BioNanoScience **8**(1), 5–16 (2018)
5. J.E. Hutchison, Greener nanoscience: a proactive approach to advancing applications and reducing implications of nanotechnology. ACS Nano **2**(3), 395–402 (2008)
6. K. Banerjee, V. Ravishankar Rai, Erratum to: a review on mycosynthesis, mechanism, and characterization of silver and gold nanoparticles. BioNanoScience **8**(1), 17–31 (2018). https://doi.org/10.1007/s12668-017-0437-8
7. G. Gahlawat, A.R. Choudhury, A review on the biosynthesis of metal and metal salt nanoparticles by microbes. RSC Adv. **9**(23), 12944–12967 (2019)
8. M. Guilger-Casagrande, R. de Lima, Synthesis of silver nanoparticles mediated by fungi: a review. Front. Bioeng. Biotechnol. **7**, 1–16 (2019)
9. L. Marchiol, A. Mattiello, F. Po, C. Giordano, R. Musetti, In vivo synthesis of nanomaterials in plants: location of silver nanoparticles and plant metabolism. Nanoscale Res. Lett. **9**(1), 101 (2014)
10. O. Collera-Zúñiga, F.G. Jiménez, R.M. Gordillo, Comparative study of carotenoid composition in three Mexican varieties of *Capsicum annuum* L. Food Chem. **90**(1), 109–114 (2005)
11. R.G. Haverkamp, A.T. Marshall, D. van Agterveld, Pick your carats: nanoparticles of gold–silver–copper alloy produced in vivo. J. Nanopart. Res. **9**(4), 697–700 (2007)
12. J. Mittal, A. Batra, A. Singh, M.M. Sharma, Phytofabrication of nanoparticles through plant as nanofactories. Adv. Nat. Sci. Nanosci. Nanotechnol. **5**(4), 043002 (2014)
13. B. Zheng, T. Kong, X. Jing, T. Odoom-Wubah, X. Li, D. Sun, Plant-mediated synthesis of platinum nanoparticles and its bioreductive mechanism. J. Colloid Interface Sci. **396**, 138–145 (2013)
14. S. Medda, A. Hajra, U. Dey, Biosynthesis of silver nanoparticles from *Aloe vera* leaf extract and antifungal activity against *Rhizopus sp.* and *Aspergillus sp.* Appl. Nanosci. **5**(7), 875–880 (2015)
15. A. Muthuvel, K. Adavallan, K. Balamurugan, N. Krishnakumar, Biosynthesis of gold nanoparticles using *Solanum nigrum* leaf extract and screening their free radical scavenging and antibacterial properties. Biomed. Prev. Nutr. **4**(2), 325–332 (2014)
16. Y.Y. Mo, Y.K. Tang, S.Y. Wang, J.M. Lin, H.B. Zhang, D.Y. Luo, Green synthesis of silver nanoparticles using eucalyptus leaf extract. Mater. Lett. **144**, 165–167 (2015)
17. A. Arumugam, C. Karthikeyan, A.S.H. Hameed, K. Gopinath, S. Gowri, V. Karthika, Synthesis of cerium oxide nanoparticles using *Gloriosa superba* L. leaf extract and their structural, optical and antibacterial properties. Mater. Sci. Eng. C **49**, 408–415 (2015)
18. F. Thema, P. Beukes, A. Gurib-Fakim, M. Maaza, Green synthesis of Monteponite CdO nanoparticles by *Agathosma betulina* natural extract. J. Alloy. Compd. **646**, 1043–1048 (2015)
19. P.V. Kumar, U. Shameem, P. Kollu, R. Kalyani, S. Pammi, Green synthesis of copper oxide nanoparticles using *Aloe vera* leaf extract and its antibacterial activity against fish bacterial pathogens. BioNanoScience **5**(3), 135–139 (2015)
20. M. Alagiri, S.B.A. Hamid, Green synthesis of α-Fe_2O_3 nanoparticles for photocatalytic application. J. Mater. Sci. Mater. Electron. **25**(8), 3572–3577 (2014)
21. S. Kannan, M. Sundrarajan, Green synthesis of ruthenium oxide nanoparticles: characterization and its antibacterial activity. Adv. Powder Technol. **26**(6), 1505–1511 (2015)
22. M. Sundrarajan, S. Ambika, K. Bharathi, Plant-extract mediated synthesis of ZnO nanoparticles using *Pongamia pinnata* and their activity against pathogenic bacteria. Adv. Powder Technol. **26**(5), 1294–1299 (2015)
23. S.P. Dubey, M. Lahtinen, M. Sillanpää, Tansy fruit mediated greener synthesis of silver and gold nanoparticles. Process Biochem. **45**(7), 1065–1071 (2010)
24. S. Ghosh, S. Patil, M. Ahire, R. Kitture, D.D. Gurav, A.M. Jabgunde, J. Bellare, *Gnidia glauca* flower extract mediated synthesis of gold nanoparticles and evaluation of its chemocatalytic potential. J. Nanobiotechnol. **10**(1), 17 (2012)

25. M. Noruzi, D. Zare, K. Khoshnevisan, D. Davoodi, Rapid green synthesis of gold nanoparticles using *Rosa hybrida* petal extract at room temperature. Spectrochim. Acta Part A Mol. Biomol. Spectrosc. **79**(5), 1461–1465 (2011)
26. H. Bar, D.K. Bhui, G.P. Sahoo, P. Sarkar, S.P. De, A. Misra, Green synthesis of silver nanoparticles using latex of *Jatropha curcas*. Colloids Surf. A **339**(1), 134–139 (2009)
27. C. Jayaseelan, R. Ramkumar, A.A. Rahuman, P. Perumal, Green synthesis of gold nanoparticles using seed aqueous extract of *Abelmoschus esculentus* and its antifungal activity. Ind. Crops Prod. **45**, 423–429 (2013)
28. K.P. Kumar, W. Paul, C.P. Sharma, Green synthesis of gold nanoparticles with *Zingiber officinale* extract, characterization and blood compatibility. Process Biochem. **46**(10), 2007–2013 (2011)
29. Y. Gao, Q. Huang, Q. Su, R. Liu, Green synthesis of silver nanoparticles at room temperature using kiwifruit juice. Spectrosc. Lett. **47**(10), 790–795 (2014)
30. E.C. Njagi, H. Huang, L. Stafford, H. Genuino, H.M. Galindo, J.B. Collins, S.L. Suib, Biosynthesis of iron and silver nanoparticles at room temperature using aqueous sorghum bran extracts. Langmuir **27**(1), 264–271 (2010)
31. M. Sathishkumar, K. Sneha, S. Won, C.W. Cho, S. Kim, Y.S. Yun, *Cinnamon zeylanicum* bark extract and powder mediated green synthesis of nano-crystalline silver particles and its bactericidal activity. Colloids Surf. B **73**(2), 332–338 (2009)
32. N.U. Islam, K. Jalil, M. Shahid, N. Muhammad, A. Rauf, *Pistacia integerrima* gall extract mediated green synthesis of gold nanoparticles and their biological activities. Arab. J. Chem. (2015)
33. F. Buazar, M. Bavi, F. Kroushawi, M. Halvani, A. Khaledi-Nasab, S. Hossieni, Potato extract as reducing agent and stabiliser in a facile green one-step synthesis of ZnO nanoparticles. J. Exp. Nanosci. **11**, 175–184 (2016)
34. P. Rajiv, S. Rajeshwari, R. Venckatesh, Bio-fabrication of zinc oxide nanoparticles using leaf extract of *Parthenium hysterophorus* L. and its size-dependent antifungal activity against plant fungal pathogens. Spectrochim. Acta Part A Mol. Biomol. Spectrosc. **112**, 384–387 (2013)
35. P. Khademi-Azandehi, J. Moghaddam, Green synthesis, characterization and physiological stability of gold nanoparticles from *Stachys lavandulifolia* Vahl extract. Particuology **19**, 22–26 (2015)
36. K. Krishnaswamy, H. Vali, V. Orsat, Value-adding to grape waste: green synthesis of gold nanoparticles. J. Food Eng. **142**, 210–220 (2014)
37. L. Castro, M.L. Blázquez, F. González, J.A. Muñoz, A. Ballester, Extracellular biosynthesis of gold nanoparticles using sugar beet pulp. Chem. Eng. J. **164**(1), 92–97 (2010)
38. C.C. Bonatto, L.P. Silva, Higher temperatures speed up the growth and control the size and optoelectrical properties of silver nanoparticles greenly synthesized by cashew nutshells. Ind. Crops Prod. **58**, 46–54 (2014)
39. S.N. Nisha, O. Aysha, J.S.N. Rahaman, P.V. Kumar, S. Valli, P. Nirmala, A. Reena, Lemon peels mediated synthesis of silver nanoparticles and its antidermatophytic activity. Spectrochim. Acta Part A Mol. Biomol. Spectrosc. **124**, 194–198 (2014)
40. H.A. Salam, P. Rajiv, M. Kamaraj, P. Jagadeeswaran, G. Gunalan, R. Sivara, Plants: green route for NPs synthesis. Int. Res. J. Biol. Sci. 85–90 (2012)
41. J.Y. Song, B.S. Kim, Rapid biological synthesis of silver nanoparticles using plant leaf extracts. Bioprocess Biosyst. Eng. **32**(1), 79–84 (2009)
42. M. Vanaja, G. Gnanajobitha, K. Paulkumar, S. Rajeshkumar, C. Malarkodi, G. Annadurai, Phytosynthesis of silver nanoparticles by *Cissus quadrangularis*: influence of physicochemical factors. J. Nanostruct. Chem. **3**(1), 1–8 (2013)
43. A. Rai, A. Singh, A. Ahmad, M. Sastry, Role of halide ions and temperature on the morphology of biologically synthesized gold nanotriangles. Langmuir **22**(2), 736–741 (2006)
44. G. Sangeetha, S. Rajeshwari, R. Venckates, Green synthesis of zinc oxide nanoparticles by *Aloe barbadensis* miller leaf extract: structure and optical properties. Mater. Res. Bull. **46**(12), 2560–2566 (2011)

45. S. Pandey, G. Oza, M. Vishwanathan, M. Sharon, Biosynthesis of highly stable gold nanoparticles using *Citrus limone*. Ann. Biol. Res. **3**(5), 2378–2382 (2012)
46. N.A. Adamafio, Theobromine toxicity and remediation of cocoa by-products: an overview. J. Biol. Sci. **13**(7), 570–576 (2013)
47. S.Y. Chan, W.S. Choo, Effect of extraction conditions on the yield and chemical properties of pectin from cocoa husks. Food Chem. **141**(4), 3752–3758 (2013)
48. A.A. Karim, A. Azlan, A. Ismail, P. Hashim, S.S.A. Gani, B.H. Zainudin, N.A. Abdullah, Phenolic composition, antioxidant, anti-wrinkles and tyrosinase inhibitory activities of cocoa pod extract. BMC Complement. Altern. Med. **14**(1), 1 (2014)
49. S.Y. Zhang, J. Zhang, H.Y. Wang, H.Y. Chen, Synthesis of selenium nanoparticles in the presence of polysaccharides. Mater. Lett. **58**, 2590–2594 (2004)
50. N. Singh, P. Saha, K. Rajkumar, J. Abraham, Biosynthesis of silver and selenium nanoparticles by *Bacillus* sp. JAPSK2 and evaluation of antimicrobial activity. Der Pharm. Lett. **6**, 175–181 (2014)
51. K.B. Narayanan, N. Sakthivel, Biological synthesis of metal nanoparticles by microbes. Adv. Coll. Interface Sci. **156**, 1–13 (2010)
52. K.S. Prasad, H. Patel, T. Patel, K. Patel, K. Selvaraj, Biosynthesis of Se nanoparticles and its effect on UV-induced DNA damage. Colloids Surf. B **103**, 261–266 (2013)
53. A. Husen, K.S. Siddiqi, Plants and microbes assisted selenium nanoparticles: characterization and application. J. Nanobiotechnol. **12**(1), 28 (2014)
54. P.B. Ezhuthupurakkal, L.R. Polaki, A. Suyavaran, A. Subastri, V. Sujatha, C. Thirunavukkarasu, Selenium nanoparticles synthesized in aqueous extract of *Allium sativum* perturbs the structural integrity of Calf thymus DNA through intercalation and groove binding. Mater. Sci. Eng. C **74**, 597–608 (2017)
55. G. Sathiyanarayanan, G. Seghal Kiran, J. Selvin, Synthesis of silver nanoparticles by polysaccharide bioflocculant produced from marine *Bacillus subtilis* MSBN17. Colloids Surf. B **102**, 13–20 (2013)

Isolation of Nanocellulose Fibers (NCF) from Cocoa Pod (*Theobroma cacao L.*) via Chemical Treatment Combined with Ultrasonication

Dzun Noraini Jimat, Sharifah Shahira Syed Putra, Parveen Jamal, and Wan Mohd Fazli Wan Nawawi

Abstract Cellulose is commonly known for its uses in the production of products and materials due to its fascinating structure and properties, and it is also an almost unlimited organic polymeric raw material. In this research, cellulose nanofibers (CNF) from cocoa pod husks (*Theobroma cacao L.*) CPH is extracted using alkaline and minimal concentration of sulfuric acid and further disintegrated with ultrasonication. The FESEM result proved that fibers with nano-dimension; diameter of 20–30 nm were observed. Based on FTIR spectrum result, it is shown that the presence of peaks of around 1170–1046 and 890–670 cm^{-1} is associated with the C–O stretching and C–H bend vibrations of the cellulose pyranose ring skeletal. The range of size distribution of CNF–CPH is within 200–400 nm as shown in the particle size analysis.

Keywords Cellulose nanofibers · Cocoa pod · Alkaline · Minimal acid concentration · Ultrasonication

1 Introduction

Cocoa (*Theobroma cacao L.*) is an economically essential crop in some countries such as Nigeria, Ghana, Indonesia and even Malaysia. Its chemical compositions on dry basis (%w/w); cellulose 35.4%, hemicellulose 37%, lignin 14.7% and remaining are ash, moisture content and others [1]. These pod husks are approximately 70–75% of the whole cocoa fruit weight. Thus, each ton of cocoa fruit would generate in range of 700 and 750 kg of cocoa pod husks (*Theobroma cacao L.*) CPH [1, 2]. In the cocoa beans industries, large quantities of cocoa pod husks (CPH) are produced as wastes in cocoa plantations which often thrown away due to no market value. Furthermore, in most cases, these husks are left to rot on the cocoa plantations without being proper treated which can cause environmental problems. It could produce foul odors; besides, the rotting CPH could propagate diseases such as black pod rot [3]. Thus,

D. N. Jimat (✉) · S. S. S. Putra · P. Jamal · W. M. F. W. Nawawi
Department of Biotechnology Engineering, Kulliyyah of Engineering, International Islamic University Malaysia, P.O. Box 10, 50728 Kuala Lumpur, Malaysia
e-mail: jnoraini@iium.edu.my

A. T. Jameel and A. Z. Yaser (eds.), *Advances in Nanotechnology and Its Applications*,
https://doi.org/10.1007/978-981-15-4742-3_6

the purposes of this study are to extract cellulose from CPH via chemical treatment assisting with ultrasonication method to further fibrillating the cellulose to obtain Nanocellulose Fibers (NCF).

2 Isolation and Characterization of Nanocellulose Fibers (NCF)

Cellulose is commonly known for its use in the production of products and materials due to its fascinating structure and properties, and it is also an almost unlimited organic polymeric raw material. Cellulose is greatly functionalized and linear rigid-chain homopolymer built by repeated joining structure blocks of d-glucose [4]. There are two kinds of nanocellulose which are cellulose nanofibrils (CNF) and cellulose nanocrystals (CNC). These two nanocelluloses have unique properties including low thermal expansion coefficient, dimensional stability, outstanding reinforcing potential and transparency, low density, high aspect ratio, biodegradability, and high strength and stiffness [4, 5]. CNF are fibrils with micrometer of lengths and nanometer of widths and commonly extracted from the lignocellulosic matrix and further fibrillating to obtain microfiber bundles. Both amorphous and crystalline cellulosic regions which are present in these nanofibers can be distinguished from CNC due to its bigger aspect ratio and that CNC are prepared from fibers via acid hydrolysis which degrades amorphous regions [5]. Due to the strong hydrogen bonding between hydroxyl groups in cellulose, it is almost hard to break the crystal parts of cellulose while the amorphous parts are moderately easy to disrupt. Wood, crop residues, tunicates, bamboo, sugarcane bagasse and bacterial cellulose are some natural resources of cellulose.

In order to extract the nanocellulose in CNC and NCF forms, the amorphous parts of cellulose have to be broken down. However, a multi-stage process is required in order to extract NCF and CNC involving vigorous chemical and/or mechanical operations. The previous studies reported that the alkaline treatment is a commonly used method to remove lignin, hemicellulose, wax and oils that cover the outer surface of the fiber cell wall. Alkaline treatment also depolymerize the inherent cellulose structure, disintegrate the external cellulose microfibrils and expose small size crystallites [6]. As a result, the dissolved lignin will be separated as a liquor rich form of phenolic compounds in the process effluent [7]. Sodium hydroxide is one of the most common chemicals used in the alkaline treatment. The delignification also can be achieved by using hydrogen peroxide which also can solubilize the hemicellulose from crude samples. However, strong alkalinity may lead to undesirable cellulose degradation; thus, extra precautions are required to ensure that the hydrolysis only occurs at the fiber surface. On the other hand, the strong acid treatment may yield aggregates with more crystalline and poor aspect ratio bundles of cellulose fibrils [8], while the mechanical method involves a large shearing force to separate fibrils which includes grinder, high-pressure homogenizer and ultrasonication.

Consequently, the micron-sized cellulose fibers can gradually disintegrate into nanofibers. However, the previous studies reported that due to the complicated multi-layered structure of the plant fibers and the interfibrillar hydrogen bonds, the fibers obtained by sonication are aggregated nanofibers with a wide distribution in width [9]. Thus, this chapter describes the extraction procedure of cellulose nanofibers/ fibrils from cocoa pod husks (NCF–CPH) via chemical treatment combined with ultrasonication.

3 Experimental

3.1 Methods and Materials

Extraction method of cellulose was adapted from Achor et al. [10] with slight modifications as described in [11, 12].

3.1.1 Cocoa Pod Husks (CPH)

1. Wash and rinse the lignocellulosic substrate, cocoa pod husk (CPH) to remove any dirt.
2. Cut CPH into a smaller size and dry them at 55 °C in an oven dryer until a constant weight of CPH was achieved.
3. Grind the dried CPH at 3500 rpm by laboratory grinder MF 10 Basic IKA WERKE and filter to get similar size particle.
4. Store the prepared CPH in a closed container at room temperature prior to use.

3.1.2 Isolation of Cellulose

1. Place CPH substrate in a 1.0 L beaker and add 2% w/v of sodium hydroxide. Immerse the beaker in a water bath maintained at 80 °C for 5 h.
2. Then, after thorough washing and filtration, bleach sample with a 1:1 aqueous dilution of sodium hypochlorite for 15 min at 80 °C.
3. Filter the sample after thoroughly washing it with sufficient amount of water.
4. Then, treat the sample with 12% w/v of sodium hydroxide at 80 °C for 1 h.
5. Then vacuum filter the treated sample and washed several times prior to treatment with 1% v/v sulfuric acid (H_2SO_4) at 80 °C.
6. Finally, vacuum filter and wash again the extracted cellulose until it reaches neutral pH before being centrifuged at 10,000 rpm for 20 min.

3.1.3 Ultrasonication

The following steps are carried out as described previously in our study [12].

1. Set the amplitude of ultrasonicator (Ultrasonic Dismembrator FB-705) to 70% and use 1/8″ microtip probe.
2. Sonicate the pellet by mixing 2.0% (w/w) of solid with sterilized distilled water (which is placed in microcentrifuge) for few hours. Put the microcentrifuge in ice bath upon the ultrasonication process.
3. Then, extract out the solid structures (pellet) and the cellulose suspension (supernatant) by centrifuging the sample at 6000 rpm for 15 min.
4. Store all samples at 4 °C until further characterization and analysis.

3.1.4 Morphology Observation Using Field Emission Scanning Electron Microscopy (FESEM)

Observe the morphology of extracted cellulose from CPH using FESEM (MERLIN ZEIS). A drop of nanocellulose suspension is placed on conductive carbon tape and left to dry. Then, the tape is mounted on the aluminum studs and sputter coated with iridium for 15 s. The fiber diameter is measured using Image J, version 6, software.

3.1.5 Structural Analysis by ATR-Fourier Transform Infrared (FTIR) Spectroscopy

Attenuated total reflectance FTIR measurements are performed on a Nicolet IS50 FTIR model spectrometer (Thermo Scientific) through the single-bounce ATR accessory equipped with germanium crystal at an angle of incidence 45°. Record FTIR spectra of the samples in the range of 400–4000 cm^{-1}.

3.1.6 Particle Size Analyzer

Particle size of the synthesized NCF is determined by using particle size analyzer (Zetasizer Nano ZS/Malvern).

4 Result and Discussion

In the present study, the cellulose nanofibers (NCF) were extracted from cocoa pod (*Theobroma cacao L.*), via chemical treatment followed by ultrasonication.

4.1 Morphological Analysis by FESEM

Figure 1a shows the FESEM micrograph of the original fibers of cocoa pod husks before treatment. Figure 1b, c shows cellulose nanofibers extracted from cocoa pod husks (NCF–CPH) with magnification of x60k and (b) x100k, respectively. The size of original CPH fiber was greatly bigger with compact structure compared to the treated CPH fibers. As can be seen in Figure (b) and (c), bundles of nanofiber with the width of 10–20 nm on the surface of cellulose nanofibers. Treatment with hydrogen peroxide and further treatment with alkali might result in complete removal of the residual cementing materials (hemicellulose) from fibers as reported by the previous studies. These are due to the complex organic compounds lignin which are soluble in alkaline compound and hemicellulose which is a water-soluble polysaccharide [6, 13]. Previous studies have reported that the hemicelluloses and pectin could be removed via acid hydrolysis by breaking down the polysaccharides to simple sugars [14], subsequently destroyed the amorphous region of cellulose fibrils [15]. On the other hand, high-intensity ultrasonication was used to disturb and disperse the micro or nanofibrils bundles in the fiber cell wall through cavitation activity [15]. It showed that alkaline and acid treatment assisted with ultrasonication could produce nanofibrils of cellulose from cocoa pod husks.

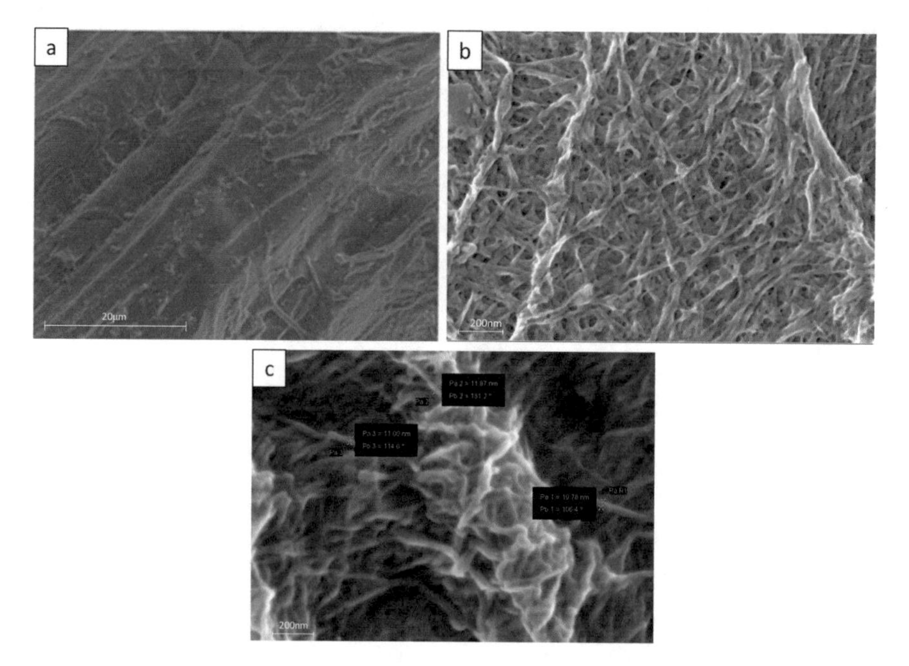

Fig. 1 FESEM micrographs of **a** original CPH without treatment (x5000); NCF–CPH extracted from cocoa pod husks at magnification of **b** x60k and **c** x100k. Reproduced from Jimat et al. [16]

4.2 Fourier Transform Infrared (FTIR) Spectroscopy

The FTIR analysis is used to detect the changes caused by the pretreatments in relation to the content of lignin and hemicellulose. After the alkali treatment, hydrogen bonding was reduced due to the removal of the hydroxyl groups by reacting with sodium hydroxide. This causes the increase of the OH^- concentration [6] as can be seen on the spectrum, a broad intensity of the peak between 3300 and 3500 cm^{-1} bands. A peak around 1645cm^{-1} corresponding to the O–H bending of water absorbed into cellulose fibers structure is also present in the treated samples. As can be seen, there is no appearance of the band at 1740cm^{-1} which is associated either to the acetyl groups in hemicelluloses or ester groups in aromatic components of lignin [17]. Furthermore, the peak at 1250 cm^{-1} that corresponds to the C–O stretching vibration of hemicellulose component of aryl-alkyl ether of lignin is also completely non-existent [17]. These results indicate that the non-cellulosic components were successfully removed from the CPH through the chemical treatment on the CPH fibres. These results indicate that the amorphous part of polymer was removed and there only remained the crystalline part of the cellulose.

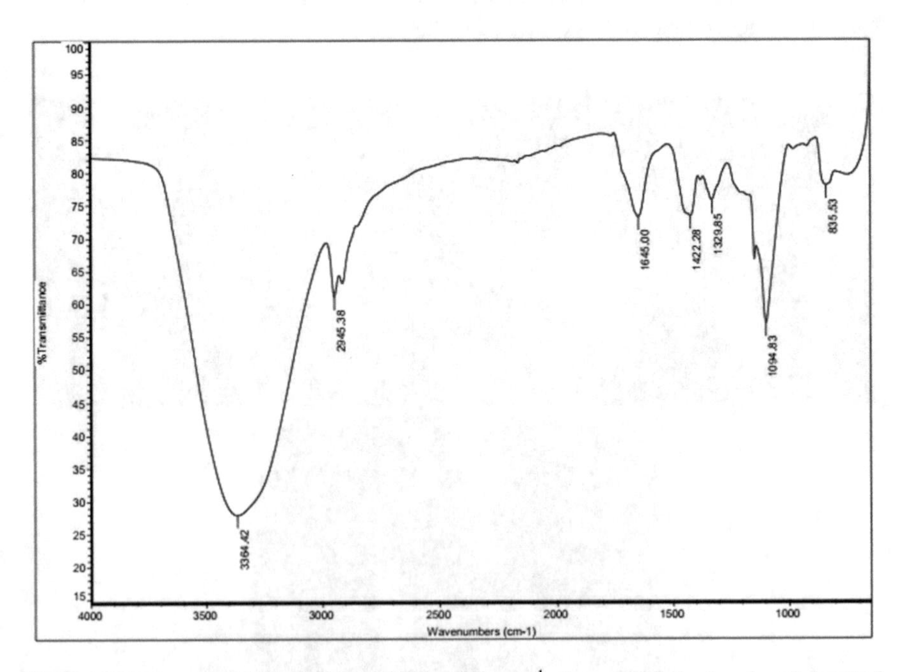

Fig. 2 FTIR spectra of NCF–CPH from 4000 to 650 cm^{-1} after going through series of chemical treatments and ultrasonication

4.3 *Particle Size Analysis*

Based on Fig. 3, it can be seen that the range of size distribution of NCF–CPH is within 200–400 nm. This result might be due to the individual cellulose fibers which were agglomerated led to increase in its dimension. Furthermore, mild sulfuric acid (1% v/v) hydrolysis might

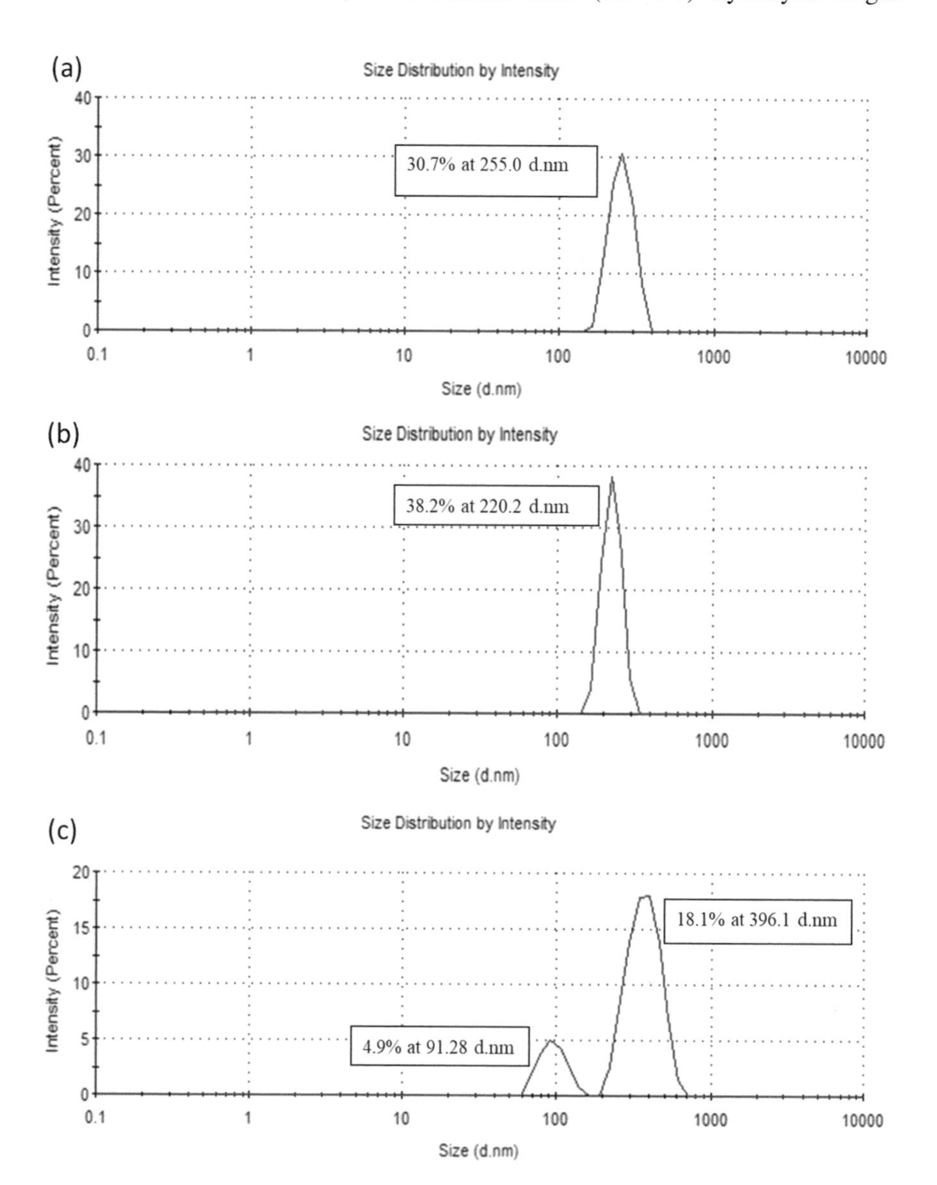

Fig. 3 Size distribution of NCF–CPH

cause the individual cellulose fibrils which were not well dispersed. The previous studies have reported that a longer acid hydrolysis (60% v/v) yielded well individual nanofibrils with lower dimension [15].

5 Conclusion

FTIR spectra results proved that the non-cellulosic components were removed after treating the CPH with chemicals. FESEM micrographs and particle size measurement confirmed that the extracted cellulose fibers were in nanoscale-sized fiber after the chemical-treated CPH samples were ultrasonicated.

Acknowledgements We acknowledge the financial support provided by the Ministry of Education of Malaysia under grant FRGS 16-044-0543.

References

1. Z. Daud, A.S.M. Kassim, A.M. Aripin, H. Awang, M.Z.M. Hatta, Chemical composition and morphological of cocoa pod husks and cassava peels for pulp and paper production. Aust. J. Basic Appl. Sci. **7**(9), 406–411 (2013)
2. G.J.F. Cruz, M. Huuhtanen, E. Alvarenga, Production of activated carbon from cocoa (*Theobroma cacao*) pod husk civil and environmental engineering production of activated carbon from cocoa (*Theobroma cacao*) pod husk. Civ. Environ. Eng. **2**(2) 2012
3. L. Cristina, R. Dias, D.M. Castanho, C. Lúcia, D.O. Petkowicz, Cacao pod husks (*Theobroma cacao L.*): composition and hot-water-soluble pectins. Ind. Crop. Prod. **34**, 1173–1181 (2011)
4. J. Kim, B.S. Shim, H.S. Kim, Y. Lee, S. Min, Review of nanocellulose for sustainable future materials. Int. J. Presicion Eng. Manuf. Green Technol. **2**(2), 197–213 (2015)
5. C. Salas, T. Nypelö, C. Rodriguez-abreu, C. Carrillo, O.J. Rojas, Current opinion in colloid and interface science nanocellulose properties and applications in colloids and interfaces ☆ fiber deconstruction. Curr. Opin. Colloid Interface Sci. **19**(5), 383–396 (2014)
6. E. Abraham et al., Extraction of nanocellulose fibrils from lignocellulosic fibres: a novel approach. Carbohydr. Polym. **86**(4), 1468–1475 (2011)
7. S.I. Mussatto, J.A. Teixeira, Lignocellulose as raw material in fermentation processes, in *Current Research, Technology and Education. Topics in Applied Microbiology and Microbial Biotechnology*, A. M.-Vilas, Ed. 2014, pp. 897–907 (January, 2010)
8. M. Paakko et al., Enzymatic hydrolysis combined with mechanical shearing and high-pressure homogenization for nanoscale cellulose fibrils and strong gels. Biomacromol **8**, 1934–1941 (2007)
9. W. Chen, H. Yu, Y. Liu, P. Chen, M. Zhang, Y. Hai, Individualization of cellulose nanofibers from wood using high-intensity ultrasonication combined with chemical pretreatments. Carbohydr. Polym. **83**(4), 1804–1811 (2011)
10. M. Achor, Y.J. Oyeniyi, A. Yahaya, Extraction and characterization of microcrystalline cellulose obtained from the back of the fruit of Lageriana siceraria (water gourd). J. Appl. Pharm. Sci. **4**(1), 57–60 (2014)
11. M. Asem, W.M.F.W. Nawawi, D.N. Jimat, Evaluation of water absorption of polyvinyl alcohol-starch biocomposite reinforced with sugarcane bagasse nanofibre: optimization using two-level

factorial design, in *IOP Conference Series: Materials Science and Engineering*, vol. 368, no. 1 (2018)
12. D.N. Jimat, Isolation of microfibrillated cellulose (MFC) via fungal cellulases hydrolysis combined with ultrasonication, in *Multifaceted Protocol in Biotechnology*, A. Azura, Ed. Singapore, pp. 109–118 (2018)
13. A. Mandal, D. Chakrabarty, Isolation of nanocellulose from waste sugarcane bagasse (SCB) and its characterization. Carbohydr. Polym. **86**(9), 1291–1299 (2011)
14. A. Alemdar, M. Sain, Isolation and characterization of nanofibers from agricultural residues—wheat straw and soy hulls. Bioresour. Technol. **99**(6), 1664–1671 (2008)
15. G.H.D. Tonoli et al., Cellulose micro/nanofibres from Eucalyptus kraft pulp: preparation and properties. Carbohydr. Polym. **89**(1), 80–88 (2012)
16. D.N. Jimat, S. Sulaiman, A.N. Yusilawati, M.A. Nor Fadhillah, S.P.S. Shahira, Physicochemical characteristics of bionanocomposites, polycaprolactone/starch/cocoa pod husk microfibrillated cellulose. J. Adv. Res. Fluid Mech. Thermal Sci **55**(2), 199–208, ISSN 2289–7879 s (2019)
17. G. Mondragon et al., A common strategy to extracting cellulose nanoentities from different plants. Ind. Crops Prod. **55**, 140–148 (2014)

Characterization of Polylactic Acid/Organoclay Nanocomposites

Fathilah Binti Ali, Azlin Suhaida Azmi, Hazleen Anuar, and Jamarosliza Jamaluddin

Abstract Polylactic acid (PLA) is a biodegradable polymer that can be substituted in usage of conventional non-degradable polymers. The strength of the PLA can be enhanced by incorporating clay. In this research, PLA was mixed with organoclay (Cloisite 30B, C30B) using Haake internal mixer. The X-ray diffraction (XRD) patterns and transmission electron microscopy (TEM) images showed that PLA chains were intercalated into the layers of the organoclay. The thermal properties showed a slight decrease in the glass transition temperature of PLA, and similar trend was observed in Tan delta. The nanocomposites exhibited higher storage modulus than the matrix. The PLA/C30B nanocomposites exhibited increment in modulus of PLA by using organoclay, in which it is suitable to replace non-degradable materials.

Keywords Polylactic acid · Filler · Organoclay · Biodegradable polymer · Environmental friendly · Nanocomposite

1 Introduction

Most of the materials used in packaging materials are made of non-degradable materials which had sparked tremendous environmental problems. To overcome the problem caused by these non-degradable materials, researchers need to find substitute materials to produce packaging materials that are biodegradable, have good

F. B. Ali (✉) · A. S. Azmi
Department of Biotechnology Engineering, Kulliyyah of Engineering, International Islamic University Malaysia, 53100 Kuala Lumpur, Malaysia
e-mail: fathilah@iium.edu.my

H. Anuar
Department of Manufacturing and Materials Engineering, Kulliyyah of Engineering, International Islamic University Malaysia, 53100 Kuala Lumpur, Malaysia

J. Jamaluddin
Department of Bioprocess and Polymer Engineering, Faculty of Chemical and Energy Engineering, Universiti Teknologi Malaysia, 81310 Johor Bahru, Malaysia

A. T. Jameel and A. Z. Yaser (eds.), *Advances in Nanotechnology and Its Applications*,
https://doi.org/10.1007/978-981-15-4742-3_7

gas permeability and not containing toxic compounds. Among the degradable polymers, polylactic acid (PLA) is an aliphatic polyester which has been investigated for packaging materials.

PLA is polymerized from its monomer which is lactic acid (LA). LA can be produced from the fermentation of starchy materials such as corn starch. Cargill Dow has been producing PLA since it is a promising environment friendly material suitable for substituting conventional polystyrene and polypropylene [1, 2]. PLA has high strength, high modulus, biodegradability and easy processing, but its brittleness and high price has limited its applications.

To improve the brittleness of PLA, blending method can also be used where PLA is blended with other flexible type polymers such as polycaprolactone [3], polybutylene succinate [4] and polyurethanes [5, 6]. PLA was also mixed with low molecular weight plasticizers such as epoxidized soybean oil [7], epoxidized palm oil [8], citrate esters [9] and low molecular weight polypropylene glycol [10].

Clay has also been studied as a filler for PLA, and the properties of the nanocomposites were reported [11–14]. The presence of clay could lead to various structural forms of clays such as intercalated, exfoliated and mixed in the nanocomposites. For this reason, the filler could enhance the properties of the nanocomposites based on its structures. These nanocomposites showed enhancement of materials properties such as crystallization rate [15], strength [16] and modulus [17].

In this research, polylactic acid (PLA) was melt blended in an internal mixer, and then, organoclay (Cloisite 30B, C30B) was incorporated into the molten polymer. The dispersion of clay, morphology, thermal and dynamic mechanical properties was studied. The results showed improved properties of PLA/organoclay nanocomposites (PLACNs), and this finding enhanced its potential use in various packaging materials.

2 Experimental

2.1 Materials

Polylactic acid and organoclay were used in this study, and the properties of the materials are listed in Table 1.

Table 1 Properties of materials

Ingredients	Description	Source
PLA	Poly(lactic acid) (average Mw = 148,000 g/mol)	Cargill Dow, USA
Organoclay	Cloisite 30B (C30B) Methyl tallow bis(2-hydroxyethyl) quaternary ammonium (where T = tallow (~65% C18; ~30% C16; ~5% C14))	Southern Clay Inc., USA

Table 2 Formulation and sample code of PLA/ESO/organoclay nanocomposites

Sample codes	PLA/organoclay (phr)
PLA	100/0
PLACN-1	100/1
PLACN-3	100/3
PLACN-5	100/5
PLACN-7	100/7

2.2 Preparation of PLA/Organoclay Nanocomposites (PLACNs)

Organoclay (Cloisite 30B) was dried at 100 °C for overnight, whereas PLA was dried at 40 °C for overnight. And then PLA was melted in an internal Haake mixer for 5 min before inserting organoclay and mixed for another 10 min. Table 2 shows the blending formulation for PLACNs. The nanocomposites were pre-melted for 10 min at 180 °C with 1 mm-thickness sheet under a pressure of 45 kN.

3 Characterization

3.1 X-Ray Diffraction (XRD)

The X-ray diffraction (XRD) patterns were recorded on a Philips, X'pert PRO MRD diffractometer using a CuKα radiation ($\lambda = 1.54$ Å) at a generator voltage of 40 kV and a generator current of 25 mA. The purpose of using XRD was to investigate the dispersibility of the organoclay silicate layers in the polymer matrix.

3.2 Transmission Electron Microscopy (TEM)

Transmission electron microscopy (TEM) was using LEO-912AB OMEGA (KBSI-CHUNCHEON) with a 120 kV accelerating voltage. The sample was microtomed before the observation. The TEM was used to observe the dispersion of the organoclay silicate layers in the polymer matrix.

3.3 Differential Scanning Calorimetry (DSC)

Differential scanning calorimetry (DSC) was conducted on a DSC2010 from TA Instruments. The DSC scans of the samples were conducted from 0 to 200 °C

(10 °C/min) and kept at 200 °C for 5 min and subsequently cooled to 0 °C (10 °C/min). The samples were then heated again to 200 °C at 10 °C/min. This second heating process was regarded as melting scan for the analysis. The crystallization and melting temperatures were determined as the temperature at the maximum values of crystallization and melting peaks. Glass transition temperature was also obtained from the second heating scan.

3.4 Dynamic Mechanical Analysis (DMA)

The dynamic mechanical properties were determined by using a dynamic mechanical analyzer (TA instrument DMA 2980). The samples were conducted in dual-cantilever bending mode with amplitude of 0.2% at a frequency of 1 Hz. The temperature was increased at a heating rate of 2 °C/min from 0 to 160 °C.

4 Results & Discussion

4.1 Dispersion of Organoclay

XRD is used to observe the dispersibility of clay platelets in the polymer. The XRD patterns of the PLA/organoclay nanocomposites are shown in Fig. 1. The peak related to the organoclay (C30B) appears at 4.74° (d_{001} spacing is 1.9 nm). The diffraction peaks of PLA nanocomposites were found to be shifted to lower angles compared

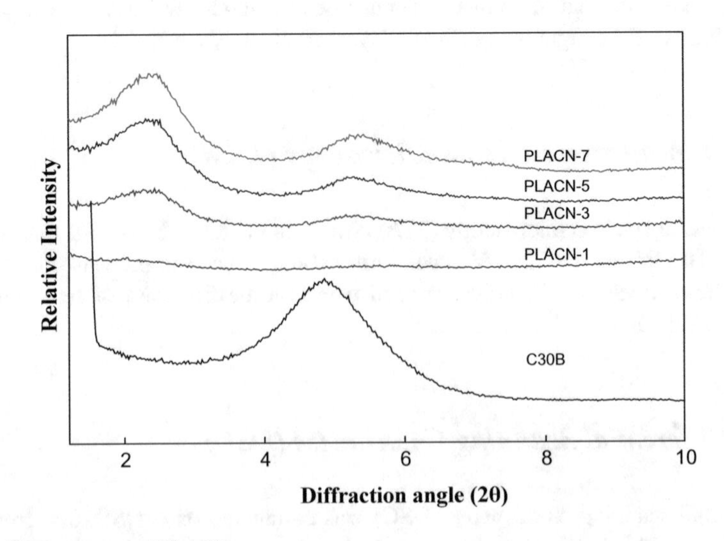

Fig. 1 XRD patterns of PLACNs

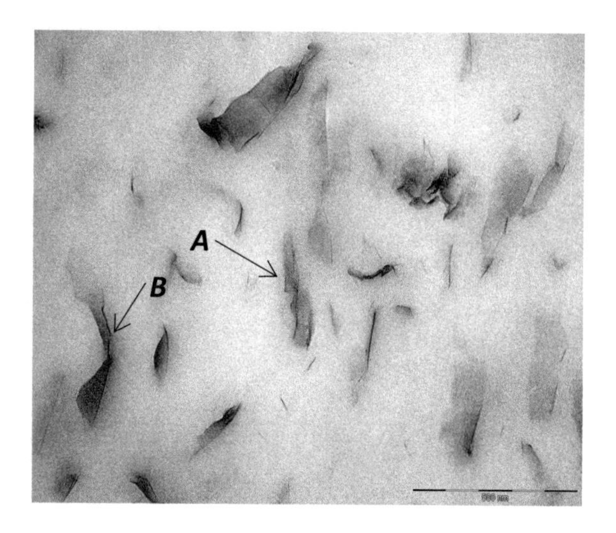

Fig. 2 TEM images of PLACN-5 (500 nm scale)

to organoclay. This indicated that PLA chains were intercalated into the layers of the organoclay [12]. Dispersion of the clay layers in the polymer matrices (PLACN-5) was confirmed in TEM (Fig. 2). Label "A" showed the silicate layers are in intercalated and in agglomerated states (label "B").

4.2 Thermal Properties

The DSC thermograms obtained from the second heating scan of PLA/clay nanocomposites (PLACNs) are shown in Fig. 3 and summarized in Table 2. The effect of organoclay on the glass transition temperature (T_g), cold crystallization temperature (T_{cc}) and melting temperature (T_m) in PLACNs was recorded. The T_g decreased with the increase in clay content, varying from 60.4 °C for PLA to 58.7 °C for PLACN-7. The T_{cc} and T_m also were decreased with clay contents. This indicated that incorporation of clay increased the free volume in the PLA matrix (Table 3) [17].

4.3 Dynamic Mechanical Properties

The temperature dependencies of storage moduli (E') and Tan δ of PLA nanocomposites are shown in Fig. 4a, b, respectively. In Fig. 4a, the nanocomposites exhibited higher storage modulus than the matrix. The storage modulus of the nanocomposites increased with the filler content [17]. Figure 4b shows the temperature dependencies of the Tan δ for PLACNs. The T_g value was determined at the maximum peak of

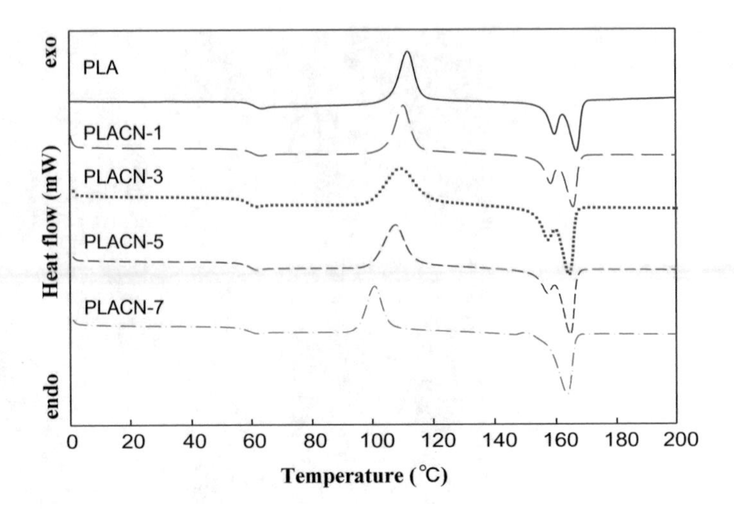

Fig. 3 DSC thermograms of PLACNs

Table 3 Thermal properties of PLACNs

Sample	T_g (°C)	Cold crystallization		Melting			
		T_{cc} (°C)	ΔH_{cc} (J/g)	T_{m1} (°C)	T_{m2} (°C)	ΔH_{m2} (J/g)	Δh_{mtotal} (J/g)
PLA	60.4	111.9	39.5	159.5	166.9	30.2	39.7
PLACN-1	60.1	110.3	34.7	158.4	165.7	27.9	35.3
PLACN-3	59.8	109.0	35.3	157.4	164.1	30.7	35.7
PLACN-5	59.4	107.6	34.9	157.4	164.7	31.0	35.4
PLACN-7	58.7	100.6	30.5	–	163.6	31.1	31.1

Tan δ. The T_g reduced slightly with addition of organoclay, and similar behavior was also observed in T_g obtained from DSC.

5 Conclusion

The PLA/organoclay nanocomposites were successfully prepared using melt mixing method. For PLACN-7, glass transition temperature, cold crystallization temperature and melting temperature were reduced about 1.7 °C, 11.3 °C, and 3.3 °C, respectively. TEM images showed that clay layers were both intercalated and agglomerated in the polymer matrices. This trend confirmed the intercalation of the organoclay by XRD. The dynamic mechanical analysis showed that the organoclays improved the modulus of the nanocomposites.

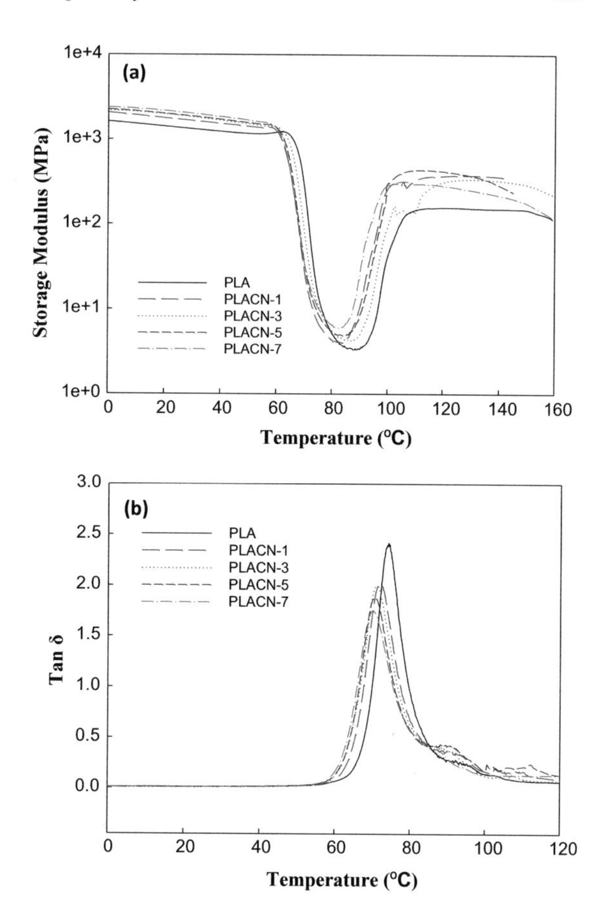

Fig. 4 Temperature dependencies of **a** storage modulus (E') and **b** Tan δ for PLA and PLACNs

Acknowledgements This work was financially supported by IIUM Research Acculturation Grant Scheme (IRAGS) 2018 (IIUM/504/RES/G/14/3/3/2) from International Islamic University Malaysia.

Conflict of Interest
The authors of this chapter have no conflict of interest.

References

1. R. Auras, B. Harte, S. Selke, An overview of polylactides as packaging materials. Macromol. Biosci. **4**, 835–864 (2004). https://doi.org/10.1002/mabi.200400043
2. R.E. Drumright, P.R. Gruber, D.E. Henton, Polylactic acid technology. Adv. Mater. **12**, 1841–1846 (2000). https://doi.org/10.1002/1521-4095(200012)12:23%3c1841:AID-ADMA1841%3e3.0.CO;2-E

3. E.A. Jaffar Al-Mulla, N.A.B. Ibrahim, K. Shameli, M. Bin Ahmad, W.M. Zin Wan Yunus, Effect of epoxidized palm oil on the mechanical and morphological properties of a PLA-PCL blend. Res. Chem. Intermed. **40**, 689 (2014). https://doi.org/10.1007/s11164-012-0994-y
4. R. Homklin, N. Hongsriphan, Mechanical and thermal properties of PLA/PBS continuous blends adding nucleating agent. Energy Procedia. **34**, 871–879 (2013). https://doi.org/10.1016/j.egypro.2013.06.824
5. W. Lin, J.P. Qu, Enhancing impact toughness of renewable poly(lactic acid)/thermoplastic polyurethane blends via constructing cocontinuous-like phase morphology assisted by ethylene-methyl acrylate–glycidyl methacrylate copolymer. Ind. Eng. Chem. Res. **58**, 10894–10907 (2019). https://doi.org/10.1021/acs.iecr.9b01644
6. J. Zhao, X. Li, H. Pan, X. Ai, H. Yang, Rheological, thermal and mechanical properties of biodegradable poly(lactic acid)/poly(butylene adipate-co-terephthalate)/poly(propylene carbonate) polyurethane trinary blown films. Polym. Bull. 1–24 (2019). https://doi.org/10.1007/s00289-019-02942-5
7. F. Ali, Y.-W. Chang, S.C. Kang, J.Y. Yoon, Thermal, mechanical and rheological properties of poly (lactic acid)/epoxidized soybean oil blends. Polym. Bull. **62**, 91–98 (2009). https://doi.org/10.1007/s00289-008-1012-9
8. F.B. Ali, R.J. Awale, H. Fakhruldin, H. Anuar, Plasticizing poly(lactic acid) using epoxidized palm oil for environmental friendly packaging material. Malays. J. Anal. Sci. **20**, 1153–1158 (2016). https://doi.org/10.17576/mjas-2016-2005-22
9. N. Ljungberg, B. Wesslén, Thermomechanical film properties and aging of blends of poly(lactic acid) and malonate oligomers. J. Appl. Polym. Sci. **94**, 2140–2149 (2002). https://doi.org/10.1002/app.21100
10. Z. Kulinski, E. Piorkowska, K. Gadzinowska, M. Stasiak, Plasticization of poly(L-lactide) with poly(propylene glycol). Biomacromolecules **7**, 2128–2135 (2006). https://doi.org/10.1021/bm060089m
11. O.H. Geun, A.K. Hyun, L.S. Jong, Effect of electric field on polymer/clay nanocomposites depending on the affinities between the polymer and clay. J. Appl. Polym. Sci. **133**, 43582 (2016). https://doi.org/10.1002/app.43582
12. M. Pluta, M.-A. Paul, M. Alexandre, P. Dubois, Plasticized polylactide/clay nanocomposites. I. The role of filler content and its surface organo-modification on the physico-chemical properties. J. Polym. Sci. Part B Polym. Phys. **44**, 299–311 (2006). https://doi.org/10.1002/polb.20694
13. B. Şengül, R.M.A. El-abassy, A. Materny, N. Dilsiz, Poly(lactic acid)/organo-montmorillonite nanocomposites: synthesis, structures, permeation properties and applications. Polym. Sci. Ser. A. **59**, 891–901 (2017). https://doi.org/10.1134/S0965545X17060098
14. S. Sinha Ray, P. Maiti, M. Okamoto, K. Yamada, K. Ueda, New polylactide/layered silicate nanocomposites. 1. preparation, characterization, and properties. Macromolecules **35**, 3104–3110 (2002). https://doi.org/10.1021/ma011613e
15. M. Pluta, A. Galeski, M. Alexandre, M.A. Paul, P. Dubois, Polylactide/montmorillonite nanocomposites and microcomposites prepared by melt blending: structure and some physical properties. J. Appl. Polym. Sci. **86**, 1497–1506 (2002). https://doi.org/10.1002/app.11309
16. M. Shibata, Y. Someya, M. Orihara, M. Miyoshi, Thermal and mechanical properties of plasticized poly(L-lactide) nanocomposites with organo-modified montmorillonites. J. Appl. Polym. Sci. **99**, 2594–2602 (2006). https://doi.org/10.1002/app.22268
17. V. Krikorian, D.J. Pochan, Poly (L-lactic acid)/layered silicate nanocomposite: fabrication, characterization, and properties. Chem. Mater. **15**, 4317–4324 (2003). https://doi.org/10.1021/cm034369+

Embedding Nano-adsorbents Within Gross Pollutant Traps (GPTs): A Review

Nurliyana Nasuha Safie, Mariani Rajin, S. M. Anisuzzaman, Mohd Zulhisham Moktar, Mohd Hazman Saafie, and Abu Zahrim Yaser

Abstract Gross pollutants (GPs) in storm water runoff have increased the number of harmful contaminants as a result of complete and incomplete decomposition of organic and non-organic matters. The addition of adsorbents may enhance the performance of GPTs as an on-site treatment. However, there is no complete review reporting the performance of GPTs with adsorbents. Hence, this review presents the preliminary process of GPTs installation, performance of GPTs in dissolved pollutants removal as well as the potential of zeolite and chitosan as adsorbent media within GPTs.

Keyword Nano-adsorbents · Zeolite · Chitosan · Gross pollutants · Traps

1 Introduction

The rapid urbanization process has caused the declined in storm water quality due to the uncontrolled and improper waste disposal. Wash-off solid waste termed as GPs is defined as discarded materials larger than 5 mm which include litter and debris as well as coarse sediment particles with grain sizes greater than 0.5 mm [1, 2]. Plastics are recorded to be the highest materials collected in the drain during monsoon, and not to mention, about 5.25 trillion plastic particles pollute the ocean surfaces [3–5]. The concern is raised when these materials are degraded by the biological, photo- and/or mechanical breakdown that reduced its size to hazardous microplastics [4, 5]. The decomposition of vegetation and food waste in storm water may increase the concentration of nutrients total phosphorus (TP) and total nitrogen (TN) [6]. Another concerning issue is the significant level of non-degradable heavy metal concentration such as Cd, Cu, Pb, Ni, Cr and Zn in urban storm water [7, 8]. Hence, this review provides a review on the common process of GPTs installation, performance of GPTs embedded with adsorbents in dissolved pollutants removal as well as the potential of zeolite and chitosan as adsorbent media in GPTs.

N. N. Safie · M. Rajin (✉) · S. M. Anisuzzaman · M. Z. Moktar · M. H. Saafie · A. Z. Yaser
Chemical Engineering Programme, Faculty of Engineering, Universiti Malaysia Sabah, Jalan UMS, 88400 Kota Kinabalu, Malaysia
e-mail: mariani@ums.edu.my

A. T. Jameel and A. Z. Yaser (eds.), *Advances in Nanotechnology and Its Applications*,
https://doi.org/10.1007/978-981-15-4742-3_8

2 Gross Pollutant Traps (GPTs)

The GPTs can be divided into two types: conventional and proprietary. Conventional GPT is a common trap installed at any type of catchment area, while proprietary GPTs are customed according to specific site characteristics [9]. The common parameters prior to installation of GPTs are the testing period, dimension of the GPTs, average rainfall depth, flow rate profile, type of pollutants and retention capacities. The performance investigation also requires preliminary studies and data collection on the catchment type and size, the pollutant load, hydrological data and runoff parameters. There are various types of GPTs such as portable rubbish and sediment trap, boom rubbish trap, CDS, Humegard®, catch basin insert (CBI), Ultra-Urban™, FloGard Plus™, HydroKleen™, Stream Guard™, Passive Skimmer, Silt Sack® and StreamGuard™. Performance of GPTs is highly depending on the installed location, population on the installed area, land use, hydrological regime, maintenance frequency, maintenance cost and specific designs [10, 11].

3 Performance of GPTs Embedded with Adsorbents in Dissolved Pollutants Removal

Based on Table 1, the number of GPTs being installed will influence the removal of dissolved pollutants such as total suspended solids (TSS), TN, TP, faecal coliform (FC), oil and grease (O&G), biological oxygen demand (BOD) and chemical oxygen demand (COD) as shown in Table 1.

Sidek et al. [12] and Alam et al. [13] recorded the highest removal of dissolved pollutants; COD and BOD due to many GPTs were installed in these two studies. Alam et al. [13] used CBI which is equipped with the filter that enables the removal of small-size contaminants. For normal GPTs such as CDS, it is specifically designed to remove GPs instead of sediments or dissolved contaminants and it is not equipped with filter or adsorbent pouch. This is also supported by the founding by Rahmat et al. [3] whereby a litter track with oil and grease trap was installed by using clay as the adsorbent has recorded 94% removal of oil and grease.

A study made by Nichols and Lucke [14] who used manufactured GPT (Humegard®) showed removal of TP, TN and TSS even though the values did not meet the water quality requirement as shown in Table 1. A combination of BMPs such as GPT, bio-retention and pond has been done by Sahrma et al. [16] increased the removal of TSS, but it requires extra space and costly due to the multiple installations of BMPs. From Table 1, it can be concluded that the number of GPTs installed and the combination of GPTs with adsorbents may enhance the removal for dissolved pollutants.

Table 1 Dissolved pollutants removal by GPTs and combination with other BMPs

Type of trap	TSS (%)	TN (%)	TP (%)	FC (%)	O&G (%)	BOD (%)	COD (%)	References
GPT	2	NA	NA	NA	NA	2	2	[12]
GPT	49	27	27	NA	NA	NA	NA	[14]
CDS	28	10	4	43	NA	NA	NA	[15]
GPT + pond	59	26	25	NA	NA	NA	NA	[16]
GPT + pond + b.rentention systems	71	16	42	NA	NA	NA	NA	
Litter trap	NA	NA	NA	NA	94	NA	NA	[3]
UST CBI	3	NA	2	NA	NA	3	3	[13]
Vortechs	50	NA	28	NA	NA	44	NA	[17]
Siltsack	47	NA	–	NA	NA	48	NA	
FloGard®	81	NA	50	NA	NA	56	NA	
HydroKleen™	85	NA	31	NA	NA	69	NA	
AbTechUltra	96	NA	84	NA	NA	68	NA	
Stream Guard™ passive skimmer	90	NA	76	NA	NA	54	NA	

NA—Not available

4 Zeolite and Chitosan as the Potential Nano-Adsorbents in GPTs

Nano-adsorbents have nanoscales pores, high selectivity, high surface area, high permeability, good mechanical stability and good thermal stability [18]. Nano-adsorbents in GPTs mostly being embedded within the GPT in capsules, pillow liner, adsorbent pillows attached to floating booms, bag filter, filter rack and adsorbent socks or being embedded freely at the bottom of GPTs [3, 17, 19]. The media fed to the system functions to target sediment-bound pollutants, metals, nutrients, hydrocarbons, grass clippings, oil and grease [13, 17, 19]. The types of adsorbents that have been used are clay [3], activated carbon, basalt [20], encapsulating polymer, sand and CPZ Mix™.

There are many natural materials that can be used as options to be used as adsorbent media in GPTs such as zeolite, chitosan, activated carbon derived from tarap skin, rice husk, Carica papaya seeds, durian (*Durio zibethinus Murray*) skin, Typha orientalis leaves, coconut shell and oil palm shell as well as an empty fruit bunch (EFB) which have been proven to treat ammonia nitrogen and other contaminants [21–26]. Activated carbon is 20 times more expensive than biochar due to the higher production cost, but it has higher removal capacity as compared to normal biochar [27]. On the other hand, EFB has been used as adsorbent which acts as a natural filter to treat ammonia nitrogen in the urban storm water which is also potential adsorbent that can be fed in GPTs [28–30].

Natural zeolite can attract contaminants due to its three-dimensional frameworks of aluminosilicates where oxygen, aluminium and silicon are covalently bonded in a tetrahedral structure which enables the ion-exchange capacity [31, 32]. The particle size of natural zeolite range between 1.0 mm [33], 2 and 20 μm [34] and nano-sized particle 550 nm [35] depending on the type of zeolite, and it is considered to be microporous materials due to its pore size normally less than 2 nm [36]. Natural zeolite has many types and has been used to treat nutrients runoff such as natural Chinese (Chende) zeolite [37], Yemeni natural zeolite [38], natural modified clays [39], coal and fly ash [40], clinoptilolite [41], Australian natural zeolite [42, 43] and Turkish clinoptilolite [44]. There are more than fifty species of natural zeolites, such as kaolin group (kaolinite, halloysite), bentonite (mostly comprising montmorillonite), heulandite, mordenite, erionite, palygorskite, sepiolite and others [36, 45, 46]. Compared to other ion-exchange materials such as organic resins, the use of natural zeolites is expedient in that they provide low-cost treatment, exhibit excellent selectivity at low temperatures, release non-toxic exchangeable cations to the environment, provide simple operation, easy maintenance at full-scale applications, compact in size and can be used in relatively little space.

Meanwhile, chitosan is a non-toxic natural carbohydrate polymer derived from the chitin component of crustacean exoskeletons such as shrimp, crab and crawfish which are reported to be an excellent adsorbent material because it contains hydroxyl (–OH) and amino ($-NH_2$) groups that serve as binding sites and reported as adsorbents for heavy metals and dye adsorption [47, 48]. The particle size of chitosan is highly related to the molecular weight, degree of deacetylation (DD) and the raw material in chitosan preparation which determines its performance as adsorbents in wastewater treatment [49, 50]. The size of chitosan may vary such as 205 nm chitosan derived from crab shell [51], 331 nm from shrimp shell [52] and 1.1 nm synthesized from fish scales [53]. Chitosan and zeolite have been extensively being used to treat contaminants that commonly present in urban stormwater and runoff such as ammonia nitrogen [54, 55], phosphate [56], dye and heavy metal [50, 57].

5 Conclusion

GPTs have raised considerable attention in reducing GPs. Incorporation of adsorbents may enhance the performance of GPTs since it can treat a high range of molecular size and different types of pollutants. Zeolite and chitosan have the potential to be adsorbents incorporated in GPTs due to their ion-exchange capacity.

Acknowledgements Authors would like to thank Universiti Malaysia Sabah for funding this work under grant SDK 0044 – 2018.

Conflict of Interest The authors of this chapter have no conflict of interest.

References

1. A. Ab Ghani, T.L. Lau, C. Ravikanth, N. Zakaria, C.S. Leow, M. Yusof, Flow pattern and hydraulic performance of the REDAC Gross Pollutant Trap. Flow Meas. Instrum. **22**, 215–224 (2011)
2. Department of Irrigation and Drainage, *Government of Malaysia Department of Irrigation and Drainage Urban Stormwater Management Manual for Malaysia MSMA 2nd edn.* (2012)
3. S.N. Rahmat, A.A.A.S. Abduh, A.Z.M. Ali, M.A.M. Razi, M.S. Adnan, Field performance of a constructed litter trap with oil and grease filter using low-cost materials. Int. J. Integr. Eng. **10**, 128–131 (2018)
4. M. Eriksen, L.C.M. Lebreton, H.S. Carson, M. Thiel, C.J. Moore, J.C. Borerro, F. Galgani, P.G. Ryan, J. Reisser, Plastic pollution in the world's oceans: more than 5 trillion plastic pieces weighing over 250,000 tons afloat at sea. PLoS ONE **9**, e111913 (2014)
5. R. Md Amin, E.S. Sohaimi, S.T. Anuar, Z. Bachok, Microplastic ingestion by zooplankton in Terengganu coastal waters, southern South China Sea. Mar. Pollut. Bull. **150**, 110616 (2020)
6. M.Z. Alam, A.H.M.F. Anwar, D.C. Sarker, A. Heitz, C. Rothleitner, Characterising stormwater gross pollutants captured in catch basin inserts. Sci. Total Environ. **586**, 76–86 (2017)
7. K.R. Reddy, T. Xie, S. Dastgheibi, Removal of heavy metals from urban stormwater runoff using different filter materials. J. Environ. Chem. Eng. **2**, 282–292 (2014)
8. A.P. Davis, M. Shokouhian, S. Ni, Loading estimates of lead, copper, cadmium, and zinc in urban runoff from specific sources. Chemosphere **44**, 997–1009 (2001)
9. M.S.F.M. Noor, L.M. Sidek, H. Basri, N.M. Zahari, N.F.M. Said, Z.A. Roseli, N.M. Dom, Evaluation of gross pollutant wet load in Sungai Sering, Malaysia, in *IOP Conference Series: Earth and Environmental Science*, Institute of Physics Publishing, p. 32 (2016)
10. N.M. Zahari, L.M. Sidek, H. Basri, N.F. Md Said, M.S.F. Md Noor, M. Jajarmizadeh, M.R. Zainal Abidin, N. Mohd Dom, Wet load study of gross pollutant traps; Kemensah River, Malaysia, in *IOP Conference Series: Earth and Environmental Science*, Institute of Physics Publishing, p. 32 (2016)
11. S.N.U. Munir, L.M. Sidek, S.H. Haron, N.F.M. Said, H. Basri, R. Ahmad, N.M. Dom, M.A. Ismail, Optimizing of gross pollutant trap to improve the maintenance at Sungai Bunus Malaysia, in *AIP Conference Proceedings*, American Institute of Physics Inc., p. 2030 (2018)
12. L. Sidek, H. Basri, L.K. Lee, K.Y. Foo, The performance of gross pollutant trap for water quality preservation: a real practical application at the Klang Valley, Malaysia. Desalin. Water Treat. **57**, 24733–24741 (2016)
13. M.Z. Alam, A.H.M.F.H.M.F. Anwar, A. Heitz, D.C. Sarker, Improving stormwater quality at source using catch basin inserts. J. Environ. Manage. **228**, 393–404 (2018)
14. P. Nichols, T. Lucke, Field Evaluation of the nutrient removal performance of a gross pollutant trap (GPT) in Australia. Sustainability **8**, 1–8 (2016)
15. G.F. Birch, C. Matthai, Efficiency of a continuous deflective separation (CDS) unit in removing contaminants from urban stormwater. Urban Water J. **6**, 313–321 (2009)
16. A.K. Sharma, S. Gray, C. Diaper, P. Liston, C. Howe, Assessing integrated water management options for urban developments—Canberra case study. Urban Water J. **5**, 147–159 (2008)
17. K. Kostarelos, E. Khan, N. Callipo, J. Velasquez, D. Graves, Field study of catch basin inserts for the removal of pollutants from urban runoff. Water Resour. Manag. **25**, 1205–1217 (2011)
18. S. Pandey, A comprehensive review on recent developments in bentonite-based materials used as adsorbents for wastewater treatment. J. Mol. Liq. **241**, 1091–1113 (2017)
19. R. Pitt, U. Khambhammettu, Field verification report for the Up-Flo^{TM} filter, small bus. Innov. Res. Phase **2** (2006)
20. D.P. Sounthararajah, P. Loganathan, J. Kandasamy, S. Vigneswaran, Removing heavy metals using permeable pavement system with a titanate nano-fibrous adsorbent column as a post treatment. Chemosphere **168**, 467–473 (2017)
21. N.N. Safie, A.Y. Zahrim, M. Rajin, N.M. Ismail, S. Saalah, S.M. Anisuzzaman, A.D. Rahayu, H. Huslyzam, R. Jennisha, T.T.H. Calvin, Adsorption of ammonium ion using zeolite, chitosan, bleached fibre and activated carbon. IOP Conf. Ser. Mater. Sci. Eng. **606**, 012003 (2019)

22. D. Krishnaiah, C.G. Joseph, S.M. Anisuzzaman, W.M.A.W. Daud, M. Sundang, Y.C. Leow, Removal of chlorinated phenol from aqueous solution utilizing activated carbon derived from papaya (Carica Papaya) seeds. Korean J. Chem. Eng. **34**, 1377–1384 (2017)
23. S.M. Anisuzzaman, C.G. Joseph, D. Krishnaiah, A. Bono, L.C. Ooi, Parametric and adsorption kinetic studies of methylene blue removal from simulated textile water using durian (Durio zibethinus murray) skin. Water Sci. Technol. **72**, 896–907 (2015)
24. S.M. Anisuzzaman, C.G. Joseph, W.M.A.B.W. Daud, D. Krishnaiah, H.S. Yee, Preparation and characterization of activated carbon from Typha orientalis leaves. Int. J. Ind. Chem. **6**, 9–21 (2015)
25. C.G. Joseph, S. Anisuzzaman, S. Abang, B. Musta, K.S. Quek, X.L. Wong, Adsorption performance and evaluation of activated carbon from coconut shell for the removal of chlorinated phenols in aqueous medium. Mater. Sci. **23**, 389–397 (2017)
26. S.M. Anisuzzaman, C. Joseph, D. Krishnaiah, W.M.A.W. Daud, E. Suali, F.C. Chee, Sorption potential of oil palm shell for the removal of chlorinated phenol from aqueous solution: kinetic investigation. J. Eng. Sci. Technol. **13**, 489–504 (2018)
27. S.K. Mohanty, R. Valenca, A.W. Berger, I.K.M. Yu, X. Xiong, T.M. Saunders, D.C.W. Tsang, Plenty of room for carbon on the ground: potential applications of biochar for stormwater treatment. Sci. Total Environ. **625**, 1644–1658 (2018)
28. L.N.S. Ricky, Y. Shahril, B. Nurmin, A.Y. Zahrim, Ammonia-nitrogen removal from urban drainage using modified fresh empty fruit bunches: a case study in Kota Kinabalu, Sabah, in *IOP Conference Series: Earth and Environmental Science*, Institute of Physics Publishing, p. 36 (2016)
29. A.Y. Zahrim, L.N.S. Ricky, Y. Shahril, S. Rosalam, B. Nurmin, A.M. Harun, I. Azreen, Partly decomposed empty fruit bunch fiber as a potential adsorbent for ammonia-nitrogen from urban drainage water, in *InCIEC 2014*, Springer Singapore, pp. 989–1001 (2015)
30. N. Bolong, I. Saad, J. Makinda, A. Yaser, M. Harun Abdullah, A. Ismail, *Influence of Oil Palm Empty Fruit Bunch (OPEFB) Agro-Waste Properties as Filtration Medium to Improve Urban Stormwater*, p. 78 (2016)
31. M. Rožić, Š. Cerjan-Stefanović, S. Kurajica, V. Vančina, E. Hodžić, Ammoniacal nitrogen removal from water by treatment with clays and zeolites. Water Res. **34**, 3675–3681 (2000)
32. N. Widiastuti, H. Wu, H.M. Ang, D. Zhang, Removal of ammonium from greywater using natural zeolite. Desalination **277**, 15–23 (2011)
33. Z. Milán, E. Sánchez, P. Weiland, R. Borja, A. Martin, K. Ilangovan, Influence of different natural zeolite concentrations on the anaerobic digestion of piggery waste. Bioresour. Technol. **80**, 37–43 (2001)
34. E.P. Favvas, C.G. Tsanaktsidis, A.A. Sapalidis, G.T. Tzilantonis, S.K. Papageorgiou, A.C. Mitropoulos, Clinoptilolite, a natural zeolite material: structural characterization and performance evaluation on its dehydration properties of hydrocarbon-based fuels. Microporous Mesoporous Mater. **225**, 385–391 (2016)
35. K. Ramesh, K. Reddy, R. Nandanan, Biswas, Nanostructured natural zeolite: surface area, meso-pore and volume distribution, and morphology. Commun. Soil Sci. Plant Anal. **45**, (2014)
36. M. Moshoeshoe, M.S. Nadiye-Tabbiruka, V. Obuseng, A review of the chemistry, structure, properties and applications of zeolites. Am. J. Mater. Sci. **7**, 196–221 (2017)
37. H. Huang, X. Xiao, B. Yan, L. Yang, Ammonium removal from aqueous solutions by using natural Chinese (Chende) zeolite as adsorbent. J. Hazard. Mater. **175**, 247–252 (2010)
38. A. Alshameri, A. Ibrahim, A.M. Assabri, X. Lei, H. Wang, C. Yan, The investigation into the ammonium removal performance of Yemeni natural zeolite: modification, ion exchange mechanism, and thermodynamics. Powder Technol. **258**, 20–31 (2014)
39. A.M. Awad, S.M.R. Shaikh, R. Jalab, M.H. Gulied, M.S. Nasser, A. Benamor, S. Adham, Adsorption of organic pollutants by natural and modified clays: a comprehensive review. Sep. Purif. Technol. **228**, 115719 (2019)
40. B. Zhang, D. Wu, C. Wang, S. He, Z. Zhang, H. Kong, Simultaneous removal of ammonium and phosphate by zeolite synthesized from coal fly ash as influenced by acid treatment. J. Environ. Sci. **19**, 540–545 (2007)

41. Ç. Müjgan, M. Yağiz, Ion exchange properties of natural clinoptilolite: lead-sodium and cadmium-sodium equilibria. Sep. Purif. Technol. **37**, 93–105 (2004)
42. G.J. Millar, A. Winnett, T. Thompson, S.J. Couperthwaite, Equilibrium studies of ammonium exchange with Australian natural zeolites. J. Water Process Eng. **9**, 47–57 (2016)
43. A. Cincotti, N. Lai, R. Orrù, G. Cao, Sardinian natural clinoptilolites for heavy metals and ammonium removal: experimental and modeling. Chem. Eng. J. **84**, 275–282 (2001)
44. D. Karadag, Y. Koc, M. Turan, B. Armagan, Removal of ammonium ion from aqueous solution using natural Turkish clinoptilolite. J. Hazard. Mater. **136**, 604–609 (2006)
45. L. Lin, Z. Lei, L. Wang, X. Liu, Y. Zhang, C. Wan, D.J. Lee, J.H. Tay, Adsorption mechanisms of high-levels of ammonium onto natural and NaCl-modified zeolites. Sep. Purif. Technol. **103**, 15–20 (2013)
46. C.V. Lazaratou, D.V. Vayenas, D. Papoulis, The role of clays, clay minerals and clay-based materials for nitrate removal from water systems: a review. Appl. Clay Sci. 105377 (2019)
47. X. Zhang, R. Bai, Mechanisms and kinetics of humic acid adsorption onto chitosan-coated granules. J. Colloid Interface Sci. **264**, 30–38 (2003)
48. M. Vakili, M. Rafatullah, B. Salamatinia, A.Z. Abdullah, M.H. Ibrahim, K.B. Tan, Z. Gholami, P. Amouzgar, Application of chitosan and its derivatives as adsorbents for dye removal from water and wastewater: a review. Carbohydr. Polym. **113**, 115–130 (2014)
49. E. Guibal, Interactions of metal ions with chitosan-based sorbents: a review. Sep. Purif. Technol. **38**, 43–74 (2004)
50. G. Crini, Recent developments in polysaccharide-based materials used as adsorbents in wastewater treatment. Prog. Polym. Sci. **30**, 38–70 (2005)
51. E. Rochima, S. Azhary, R. Pratama, C. Panatarani, I.M. Joni, Preparation and characterization of nano chitosan from crab shell waste by beads-milling method. IOP Conf. Ser. Mater. Sci. Eng. **193**, 12043 (2017)
52. M. Ali, M. Aboelfadl, A. Seliem, H. Khalil, G. Elkady, *Chitosan nanoparticles extracted from shrimp shells, application for removal of Fe(II) and Mn(II) from aqueous phases* (Sep. Sci, Technol, 2018)
53. I. Hermiyati, I. Iswahyuni, S. Juhana, Synthesis of chitosan from the scales of starry trigger fish (Abalistes Stelaris). Orient. J. Chem. **35**, 377–383 (2019)
54. M. Gaouar Yadi, B. Benguella, N. Gaouar-Benyelles, K. Tizaoui, Adsorption of ammonia from wastewater using low-cost bentonite/chitosan beads, Desalin. Water Treat. **57**, 21444–21454 (2016)
55. N.N. Safie, A. Zahrim Yaser, N. Hilal, Ammonium ion removal using activated zeolite and chitosan. Asia-Pacific J. Chem. Eng. 1–9 (2020)
56. X. Cui, H. Li, Z. Yao, Y. Shen, Z. He, X. Yang, H.Y. Ng, C.-H. Wang, Removal of nitrate and phosphate by chitosan composited beads derived from crude oil refinery waste: Sorption and cost-benefit analysis. J. Clean. Prod. **207**, 846–856 (2019)
57. E. Guibal, E. Touraud, J. Roussy, Chitosan interactions with metal ions and dyes: dissolved-state versus solid-state application. World J. Microbiol. Biotechnol. **21**, 913–920 (2005)

Author Index

A. T. Jameel and A. Z. Yaser (eds.), *Advances in Nanotechnology and Its Applications*,
https://doi.org/10.1007/978-981-15-4742-3

Printed in the United States
By Bookmasters